Medium-Size Photovoltaic Power Plants

Scientific editing: W. SCHNELL
Publication arrangements: D. NICOLAY

Commission of the European Communities

Medium-Size Photovoltaic Power Plants

Proceedings of an EEC/DOE Workshop
hosted by the Commissariat à l'Energie Solaire
and held in Sophia-Antipolis, France, 23-24 October, 1980

Edited by

HENRY L. DURAND
Commissariat à l'Energie Solaire (France)

PAUL D. MAYCOCK
U.S. Department of Energy

WOLFGANG PALZ
Commission of the European Communities

D. REIDEL PUBLISHING COMPANY
DORDRECHT : HOLLAND / BOSTON : U.S.A.
LONDON : ENGLAND

Library of Congress Cataloging in Publication Data

CIP

Main entry under title

Medium-size photovoltaic power plants

At head of title: Commission of the European Communities.
Includes index.
1. Photovoltaic power generation—Congresses.
I. Durand, Henry L. II. Maycock, Paul D. III. Palz, Wolfgang. IV. Commission of the European Communities.
TK2960.M42 621.31'244 81-4298
ISBN 90-277-1279-4 AACR2

Organization of the Workshop by
Commission of the European Communities
Directorate-General Research, Science and Education, Brussels

Publication arrangements by
Commission of the European Communities
Directorate-General Information Market and Innovation, Luxembourg

EUR 7090

Published by D. Reidel Publishing Company
P.O. Box 17, 3300 AA Dordrecht, Holland

Sold and distributed in the U.S.A. and Canada
by Kluwer Boston Inc.,
190 Old Derby Street, Hingham, MA 02043, U.S.A.

In all other countries, sold and distributed
by Kluwer Academic Publishers Group,
P.O. Box 322, 3300 AH Dordrecht, Holland

D. Reidel Publishing Company is a member of the Kluwer Group

Printed in The Netherlands

CONTENTS

SESSION 3 - EC PILOT PROJECTS

WELCOME ADDRESS

PHOTOVOLTAIC POWER SYSTEM WORKSHOP

W. PALZ

Commission of the European Communities

Brussels, Belgium

Mesdames, Messieurs, Ladies and Gentlemen,

I have the honour and the pleasure to open this Workshop on photovoltaic pilot projects in the medium power range (100 kW or so).

These projects are considered to be very important for the future of photovoltaics in general and for us in the European Communities they highlight the general R&D activities which are already going on in this field.

I would like to welcome here in particular our friends from the United States. I am delighted to see that it has become possible to organize this Workshop jointly between authorities in Europe and the United States and I have to acknowledge here our excellent cooperation with Paul Maycock, the Director of Photovoltaics at the D.O.E. I welcome this opportunity to demonstrate our friendship towards all our colleagues from America. I feel that cooperation between Europe and the U.S. on this topic is very important; it is interesting also to note that if we have come together for this 2 days workshop, we only follow an on-going cooperation between industry from both sides of the Atlantic.

I hope that this workshop will be very successful and that many other opportunities of this kind will arise in the future.

SESSION 1

US PROGRAMMES AND MARKETS

Chairman : P.D. MAYCOCK, US Department of Energy

- Overview of US, DOE large photovoltaic systems
- Photovoltaic stand-alone systems projects
- US Grid-connected photovoltaic application experiments
- Photovoltaic systems design comparisons between grid-connected and stand-alone systems
- Comparison of direct current and alternating current loads in a commercial building with a photovoltaic power supply

OVERVIEW OF U.S., DOE LARGE PV SYSTEMS

Paul D. MAYCOCK

Director

Photovoltaic Energy Systems Division

U.S. Department of Energy

SUMMARY

The purpose of this paper is to introduce the U.S. participants in the workshop and to summarize the status of U.S. large system experiments.

1. U.S. PARTICIPANTS

The U.S. Department of Energy is excited and pleased that the European Community asked that we participate in a workshop on Photovoltaic Energy System Designs. First I will present the persons who manage the DOE funded projects, then I will introduce the participants from the U.S. Industry who are actually making the projects a reality.

Our DOE project managers include Dr. Ed. Burgess, Dr. Ed. Kern, Dr. Tony Ratajczak and Dr. Hal Macomber who recently left DOE and joined Monegon Inc.

Dr. Macomber was the DOE manager who prepared the request for proposal and selected the contractors for the U.S. first large PV system effort. One key factor that he established was a two-step process. Dr. Macomber caused first a design phase, then a hardware phase. We had 149 proposals, selected 27 for detailed design, then we had another competition and selected 9 projects costing 22 million dollars for construction.

Dr. Ed. Burgess of Sandia National Labs. has managed all of the projects that Dr. Macomber funded. Ed has developed a detailed technical and economic management system that is working well. He will describe our status, problems, and outlook.

Dr. Tony Ratajczak is an engineering manager at the Nasa Lewis Lab. in Ohio. Tony is involved with the Indian village at Scuchulic, the village in Upper Volta, and tens of stand-alone remote projects throughout the world.

Dr. Kern is from the MIT, Lincoln Laboratories. He will describe the Natural Bridges 100 kilowatt project and our residential program.

The U.S. Industry participants will describe their experience with large PV systems. They are:

- Dr. Harry S. Zwibel, New Mexico Solar Energy Institute
- Dr. Henri J.R. Maget, Varian Associates Inc.
- Dr. Richard A. Whisnant, Research Triangle Institute
- Dr. Matthew S. Imamura, Martin Marietta Corporation
- Dr. Richard Addiss, Solar Power Corporation
- Dr. Harold L. Macomber, Monegon Ltd.

2. OVERVIEW OF U.S. PROJECTS

The figures below list and describe all of the U.S. large system projects. They will be discussed in detail by the U.S. participants. Figure 1 lists the 5 large systems that are operational, figures 2 to 4 list all DOE large system projects and show the associated U.S. utility that is supporting the project. Because of the technical and economic factors involved with our "on-the-grid" experiments, we required local utility participation. Figure 5 shows our residential projects. We are very excited about the prospects for grid-connected residences in the U.S. Our first 7 kW residence in Phoenix has created incredible interest.

I am proud of the engineering managers who have come to be with you. The next two historic days of U.S. - European collaboration will prove valuable not only to our programs but to the acceleration of photovoltaic systems deployment throughout the world.

FIGURE 1

INTERMEDIATE PROJECTS

OPERATIONAL

1. MEAD, NEBRASKA - AGRICULTURAL EXPERIMENT

2. MT. LAGUNA, CA - AFS DIESEL MINI-GRID

3. WBNO; BRYAN OH - DAYTIME RADIO STATION

4. NATURAL BRIDGES NATIONAL MONUMENT, UTAH - REMOTE VISITOR CENTER

FUTURE

THIRTEEN (13) PROJECTS : HOSPITAL, AIRPORT, SHOPPING CENTER, SCHOOLS, OFFICE BUILDINGS, PUBLIC BUILDINGS, ETC.

FIGURE 2

PHOTOVOLTAIC PROJECTS WITH UTILITY PARTICIPATION

PROJECT	SIZE(kW)	DATE OPERATIONAL	ASSOCIATED UTILITY
(PRDAs) - Flat Plate)			
Newman Power Station - El Paso, TX	20	Nov. 80	El Paso Electrc[1]
Oklahoma Center for Science & Arts - Oklahoma City, OK	150	Feb. 81	Oklahoma Gas & Electric [2]
Shopping Center - Lovington, NM	100	Dec. 80	Lea County Electric Coop. [1]
Beverly Hill School - Beverly, MA	100	Mar. 81	New England Power Service [1]

[1] Agreement made

[2] Agreement pending

FIGURE 3

PHOTOVOLTAIC PROJECTS WITH UTILITY PARTICIPATION

PROJECT	SIZE(kW)	DATE OPERATIONAL	ASSOCIATED UTILITY
(PRDAs - Concentrators)			
Sky harbor Airport - Phoenix, AZ	225	Sept. 81	Arizona Public Service[1]
Dallas - Fort Worth Airport - TX	27	June 81	Texas Power & Light [1]
BDM Office Building - Albuquerque, NM	47	Dec. 80	Public Service of NM [1]
G.N. Wilcox Hospital - Kauai, HI	60 [2]	Mar. 81	Hawaii Electric[1]
Sea World - Orlando, FL	110	Oct. 81	Florida Power Corp[1]

[1] Agreement made

[2] Pending with Acurex

FIGURE 4

PHOTOVOLTAIC PROJECTS WITH UTILITY PARTICIPATION

PROJECT	SIZE(kW)	DATE OPERATIONAL	ASSOCIATED UTILITY
(Schools)			
Mississippi County Community College - Blytheville, AK	250	Dec. 80	Ark-Mo Power [1]
Northwest Mississippi Junior College - Senatobia, MS	200	Jan. 81	Mississippi Power & Light [1] [2]
Georgetown University Campus Building - Washington, D.C.	300	Late 1984	Potomac Electric Power[1] [3]

[1] Agreement made

[2] No sell-back because no excess power

[3] Not yet formalized

FIGURE 5

RESIDENTIAL PROJECTS

° PHOENIX HOUSE (JOHN LONG, INC)	NOW
° FLORIDA SOLAR ENERGY CENTER HOUSE	8/80
° THREE HOUSES IN HAWAII	8/80
° FOUR PROTOTYPES AT THE N.E. RESIDENTIAL EXPERIMENTAL STATION	1980
° SEVERAL PROTOTYPES AT THE S.W. RESIDENTIAL EXPERIMENTAL STATION	1981
° TENS OF LIVED-IN EXPERIMENTAL HOUSES NEAR EACH RES;	1981

PHOTOVOLTAIC STAND-ALONE SYSTEMS PROJECTS

Managed for the
United States Department of Energy
and the
United States Agency for International Development
by the
NASA Lewis Research Center

by
A. F. Ratajczak
NASA-Lewis Research Center

1. INTRODUCTION

The Lewis Research Center (LeRC) of the NASA has been a leader in photovoltaic (PV) research and development since 1960. In the early 1970's PV system technology developed for the space effort was applied to terrestrial applications with the subsequent installation of several small PV systems for use by other Federal Agencies. In 1976, the LeRC was asked to manage what has become the Department of Energy's (DOE) PV Stand-Alone Applications Project. The objective of this Project is to develop PV systems for applications which represent potentially significant worldwide, near- and mid-term markets, and to demonstrate these systems cooperatively with users in a manner most likely to stimulate those markets. Also in 1976, the U.S. Agency for International Development (US/AID) requested that the LeRC design and install a PV system for water pumping and grain milling for a village in Upper volta, Africa, in support of its program "Studies of Energy Needs in the Food System." This beginning relationship with the US/AID has broadened into an extensive project to demonstrate the suitability of PV power systems for meeting basic electrical requirements in rural areas of developing countries. This work is being carried out for the US/AID Bureau for development Support, Office of Energy, to facilitate the use of Photovoltaics for a variety of applications in support of AID's development assistance activities. This paper references and briefly describes the systems the LeRC has deployed, summarizes significant findings from the system's operations, and gives some insight into the type and range of planned DOE & AID projects.

2. SUMMARY OF INSTALLED SYSTEMS

Table I summarizes LeRC-stand-alone systems by load type, date of installation, total installed watts, and installation location.

By way of background, all LeRC systems except the High Volume Air Sampler system were designed, built, and installed by LeRC personnel. Also, all LeRC stand-alone systems are all DC systems, i.e., all loads accept DC power. All-DC systems were used to enhance overall system efficiency and thus keep system costs low. To appreciate this approach, one needs to understand that there are three additional costs associated with providing AC power; the cost of the inverter itseslf, the cost of additional PV array needed to supply the inverter no-load energy requirements, and the cost of additional array to compensate for the inverter operating inefficiency. For small systems inverter losses alone can as much as double the size of the PV array over that required for a comparable all-DC system.

TABLE I - SUMMARY OF OPERATING DOE & AID
STAND-ALONE PHOTOVOLTAIC SYSTEMS

Type of System	No. of Systems	Date Installed	Total Installed Watts	Locations
Refrigerator	2	1976	550	MI, AZ
Forest Lookout Towers	2	1976	588	CA
Highway Conditions Sign	1	1977	116	AZ
Insect Survey Traps	4	1977	372	TZ
Automatic Weather Stations	6	1977	592	NY, ME, FL, NM, AK, HI, CA
Water Cooler	1	1977	446	CA
Village Power	2	1978/79	5300	AZ, AFRICA
High Volume Air Sampler	1	1979	360	NJ
Seismic Sensors	1	1980	40	HI

The refrigerator systems; forest lookout towers, highway conditions sign, weather stations, insect traps, and water cooler systems, are described elsewhere.(1) The Schuchuli and Tangaye systems are described in reference 2. A High Volume Air Sampler System is at Liberty State Park, New Jersey, directly opposite the Status of Liberty. It is a 60 VDC system and contains a 360 Wpk array and a 561 amp-hour battery. Installation was completed and the system became fully operational in May 1980. Although the high volume air sampler is part of the New Jersey Department of Environmental Protection air pollution monitoring network, the system was designed primarily to demonstrate a PV system to people from the metropoliton New York area. A seismic sensor system is located at Heiheiahula, Puna, on the Island of Hawaii. It is a 12 VDC system, contains a 40 Wpk array and a 75 amp-hour battery, and powers a tiltmeter, a seismometer, a signal processor, and a VHF transmitter. The system was installed in January 1980 and is operated by the Hawaiian Volcanoes Observatory.

3. SIGNIFICANT OPERATIONAL RESULTS

In general, PV systems have proven to be reliable and to require minimum maintenance. As a prime example, the two forest lookout tower systems have a 100% on-line record since their October 1976 installation. One system has required absolutely no maintenance even to adding water to the battery cells. The other system has required only an annual array washing. The Schuchuli and Tangaye PV systems have, respectively, 99% and 96% on-line records. Refrigerators have been a problem. Commercially available recreational vehicle type refrigerators proved inadequate in a desert environment. The special design high efficiency refrigerators at Schuchuli initially experienced motor failure problems that were eventually eliminated by replacing the 1/8 hp motors with 1/4 hp motors. There have been control system problems at both Schuchuli and Tangaye resulting from component failures (e.g., transistors, meter relay lights), but not from control systems design.

The socio-economic effects at Schuchuli have been minimal.(3) There has been little, if any, change in lifestyle beyond different hours of waking, eating the evening meal, and retiring due to the presence of electric lights. In Tangaye, however, the presence of the PV system and load services (pumped water & milled flour) have had many and varied effects.(4) The mere presence of the PV system has brought status and recognition to the village. Pumped water has affected building (making of mud bricks) and personal hygiene in the immediate area. The mill has had a complex but net positive effect on the villagers. The US/AID and the LeRC are planning to increase the PV array size to permit 48 hrs/week of mill operation in contrast to its present 20 hr/week limitation.

4. FUTURE PROJECTS

Several international stand-alone projects are now in the planning or implementation stages. The largest is the AID Tunisian Renewable Energy Resources Project which will include a PV system for a village of about 125 people, PV-powered micro-irrigation systems, and a PV system for a farmhouse. Installation is planned for 1982.

A project of importance to the world health community is the DOE/U.S. Center for Disease Control [CDC] PV-powered medical refrigerator project. Contracts for these systems will be awarded in late 1980 and will result in mid-1981 in PV-powered refrigerator systems capable of meeting both CDC and World Health Organization cold chain refrigerator requirements. An AID project will provide standardized rural health center PV systems for Ecuador, Guyana, Kenya, and Zimbabwe. These systems will be designed as standardized package systems capable of providing power for a refrigerator, lights, sterilizer, and one or two optional medical loads (e.g., dental equipment). And finally, a DOE village power systems project will result in standardized services oriented packaged PV systems (e.g., dispensay, school, pump) for four remote villages in Gabon.

5. CONCLUDING REMARKS

The experience gained from 45 system-years of PV systems powering a diversity of loads in a variety of geographic and environmental settings has shown that photovoltaics are not only suitable, but probably preferred stand-alone power systems. They are reliable, require little maintenance, and are capable of being operated by local personnel. The greatest impediments to the widespread use of photovoltaics is the high initial cost and a reluctance associated with embracing a new technology. The U.S. DOE and AID programs, although different in objective, are each in its own way aimed at overcoming these impediments.

References

1. Description and Status of NASA-LeRC/DOE Photovoltaic Applications Systems Experiments, A. F. Ratajczak; Procedings, 13th Photovoltaic Specialists Conference, Washington, DC, June 5-8, 1978.

2. Description of Photovoltaic Village Power Systems in the United States and Africa, A. F. Ratajczak and W. J. Bifano; Prepared for Photovoltaic Solar Energy Conference, Berlin, West Germany, April 23-26, 1979.

3. Social and Economic Impact of Solar Electricity at Schuchuli Village: A Status Report, W. J. Bifano and A. F. Ratajczak (NASA/LeRC), and D. M. Bahr and B. G. Garrett (Arizona State Univ.); Prepared for Seminar on Solar Technology in Rural Settings: Assessments of Field Experiences; sponsored by United Nations Univ., Atlanta, Georgia, USA, June 1-2, 1979.

4. A Final Evaluation of the Social Impact of the Tangaye (Upper Volta) Solar Energy Demonstration, A. F. Roberts, Univ. of Michigan and Office of Rural Development, U.S. AID, Upper Volta, to be published.

U.S. Grid-Connected Photovoltaic Application Experiments

E.L. Burgess

Sandia National Laboratories

Albuquerque, N.M., USA

In 1977 the U.S. Department of Energy initated a program to design, install and operate several grid-connected photovoltaic application experiments. The size range allowed for these experiments was 20 to 500 kilowatts peak. The photovoltaic arrays could be either flat panels or concentrating modules. Combined systems producing both heat and electricity were allowed; concentrator systems utilizing active cooling were particularily suited to this approach. The overall objective of this program was to gain PV system level experience with intermediate-sized PV systems operating in concert with the grid.

One hundred-fifty (150) proposals were received in response to the program announcement. Twenty-nine (29) projects were selected for a design phase; 12 projects were flat panel experiments and 17 projects proposed concentrator arrays. The total cost of this design phase was 8.4 million dollars. The design phase was completed in January 1979. Nine (9) projects were selected for fabrication and installation. Brief details of these projects are given in Table I.

Upon completion, these projects will be operated under the Department of Energy sponsorship for a period of two years. Complete performance data will be collected and analyzed by a DOE Data Reduction Center. The performance data will be distributed to interested system designers for evaluation and future design improvement.

Table I - U.S. Grid-Connected Photovoltaic Application Experiments

Prime Contractor	Location	Application	Size (kW_p)	Array Type	Initial Operation	Contract Amount
Acurex Corp.	Kauai, HI	Hospital	60	Concentrator	August,1981	$ 1,781,000
Arizona Public Service	Phoenix, AZ	Airport	225	Concentrator	November,1981	$ 5,400,000
BDM Corp.	Albuquerque, NM	Office Bldg.	50	Concentrator	February,1981	$ 1,111,000
E-Systems, Inc.	Dallas-Ft. Worth, TX	Airport	27	Concentrator	July,1981	$ 785,000
General Electric	Orlando, FL	Seaworld Park	110	Concentrator	December,1981	$ 3,386,000
Lea County Electric	Lovington, NM	Shopping Center	100	Flat Panel	January,1981	$ 2,714,000
New Mexico Univ.	El Paso, NM	Computer UPS	20	Flat Panel	November,1980	$ 476,000
Science Applications, Inc.	Oklahoma City, OK	Museum	135	Flat Panel/ Reflectors	May,1981	$ 2,700,000
Solar Power Corp.	Beverly, MA	High School	100	Flat Panel	February,1981	$ 2,874,000

PHOTOVOLTAIC SYSTEMS DESIGN COMPARISONS BETWEEN GRID-CONNECTED AND STAND-ALONE SYSTEMS

Dr. H.L. Macomber
Vice President
MONEGON, LTD.
Gaithersburg, Maryland U.S.A.

1. INTRODUCTION

Design considerations for grid-connected and stand-alone photovoltaic (PV) power systems generally will lead to not only differences in system configurations but also to differences in design methods or approaches. These differences derive from the questions of system reliability and economics and what is required analytically to answer these questions.

This paper briefly discusses key trade-offs which are considered in addressing the questions of system reliability and economics for grid-connected and stand-alone system design and illustrates these issues with two examples from MONEGON's photovoltaic system design experience.

2. SYSTEM RELIABILITY AND ECONOMICS

In examining the issue of reliability in power system design, the engineer is faced with issues of how much redundancy and/or independent backup is required to meet some specified reliability criterion. The analytical tools which are used involve statistical probability models which attempt to forecast the likelihood that there is sufficient power generation capacity to meet the load under a variety of system configurations.

In a photovoltaic power system there is the additional problem of adequate solar radiation to provide sufficient power when there is a demand. Grid-connected photovoltaic system designers usually improve system reliability by taking maximum advantage of the backup available by the utility power grid-connection. There are, however, economic and reliability trade-offs which are made between including some PV system energy storage and the backup utility power. Photovoltaic array sizing as well as other area-related components (e.g. wiring) are affected by the results of these trade-offs. In the U.S. regulated electric utility situation gird-connected PV array size trade-offs will also include the economic effect of selling any excess energy generated by the on-site PV system as required by recent regulations calling for the utilities to buy such electricity at fair equitable rates.

Simulation models which consider hour-by-hour solar radiation PV system performance, load demand and which contain economic optimization routines are generally sufficient for accurate PV system design. For example, a computer program called SOLCELL 2 which was developed by the Sandia Laboratories in Albuquerque, New Mexico is widely used for PV system design.

Stand-alone photovoltaic power systems design require somewhat different considerations than grid-connected systems. The stand-alone PV system will be either fully capable of meeting the load or the load demand will not be met. System designs, for stand-alone systems have generally addressed this reliability issue by oversizing either the PV array, the energy storage or both.

MONEGON has been developing a new method for systems design for stand-alone PV systems. This method is based upon a loss-of-load probability (LOLP) computations. The PV array size and energy storage capacity are sized to meet some defined reliability criterion. Thus, the method leads to a stand-alone PV system size which is cost-effective.

3. SYSTEM DESIGN ILLUSTRATIONS

The Solarex Corporation is being funded under a U.S. Department of Energy project to design and install a 35 kW peak photovoltaic power system for a light manufacturing facility located in San Bernardino, California, U.S.A. MONEGON, under a subcontract with Solarex, conducted a conceptual design and system analysis to select the most cost-effective photovoltaic system configuration or this grid-connected application.

The conceptual systems analyses determined the performance and economics of four PV conceptual systems sized to meet the design constraint of delivering 35 kW peak to the load on June 21. The four system configurations were: a flat-plate, fixed-tilt PV array with a power conditioner, a maximum power tracker and no battery storage; a two-axis tracking, flat-plate PV array with a power conditioner, a maximum power tracker and no battery storage; a two-axis tracking concentrating, linear focus PV array with a power conditioner, a maximum power tracker and no battery storage; and a fixed-plate, fixed-tilt PV array with a power conditioner, with no maximum power tracker and with battery storage. The last configuration was examined with three different storage capacities: one day, two days and three days of storage.

The analyses conducted established that the most cost-effective system configuration consisted of a flat-plate, fixed tilt photovoltaic array interfacing with a

Dc-Ac inverter providing power in parallel with the electric utility power. The array is tilted at a 25° angle and is sized at 332 m^2 to deliver 35 kW at the peak insolation condition on June 21. The system has no battery storage. Results from SOLCELL computer analyses for the design configurations are presented below.

Table I System Comparison

System	Annual Energy to Load (kWH)	Annual Energy Demand (kWH)	Percent Demand Met by PV	Annual Back-Up Energy Required (kWH)	Annual Excess Energy (kWH)
Fixed Flat-Plate No Battery	67841.35	207396.04	33	139554.70	21133.55
Battery (1 day)	47104.16	207396.04	23	159115.27	9774.12
Tracking Flat-Plate	81670.10	207396.04	39	126495.32	31851.56
Concentrator	73848.96	207396.04	36	134277.19	28068.63

Table II

Conceptual System	Total System Costs (1980$)		Annual Costs (1980$)				Levelized Energy Costs, (1980$) $/kWh	
	Array	BOS	Capital	Battery	O&M	Back-Up Energy	PV System	Backup
Flat-Plate, Fixed-Tilt, No Battery	318,434	142,454	53,895	—	12,493	47,222	0.489	0.169
Tracking Flat-Plate, No Battery	353,816	230,908	68,376	—	15,850	42,803	0.521	0.169
Tracking, Concentrating, No Battery	363,590	235,795	70,090	—	16,248	45,436	0.590	0.169
Flat-Plate, Fixed-Tilt, with 1 Day Battery Storage	318,434	156,562	53,369	2,175	12,876	53,841	0.709	0.169

3.2 Stand-Alone System

MONEGON's loss-of-load probability (LOLP) method for system sizing of stand-alone PV systems has been used to compute array sizes and battery capacities for village power system applications. To illustrate the method, a systems design analyses was made for the load profile developed for the Papago Indian Village Photovoltaic Power System designed and installed by the U.S. government in 1978.

The system was installed with a 3.5 kW peak photovoltaic array and a 143 kWh storage battery plus various other system components. Using the MONEGON LOLP method for a one percent LOLP it was found that the most cost-effective design size used a 3.4 kW peak photovoltaic array with a 51 kWh storage battery. Thus, the installed storage battery was oversized by a factor of nearly 3 times the analyzed size. (The actual battery capacity of 143 kWh would decrease the LOLP to approximately 0.001%.). Also, the battery capital cost would represent less than 4% of the total capital cost; whereas, the installed battery cost represents 11% of the total capital cost. Therefore, the system design method is critical for both proper performance and for achieving cost-effectiveness.

COMPARISON OF DIRECT CURRENT AND ALTERNATING CURRENT LOADS IN A COMMERCIAL BUILDING WITH A PHOTOVOLTAIC POWER SUPPLY

R. A. Whisnant
Energy Systems Department
Research Triangle Institute

1. Problem Statement

Photovoltaic (PV) systems have been shown to have considerable potential for application to the intermediate load sector typified by such establishments as retail buildings, small office buildings, schools, supermarkets, and the like (1). These are attractive applications because their daily load profiles tend to match solar insolation profiles, their electrical loads per unit floor space are relatively low in some cases, and under the conventional electricity rates applicable (i.e., declining block rates) their average cost of electricity is higher than larger industrial users.

Because photovoltaic arrays produce direct current electricity which must then be inverted to alternating current to drive the conventional loads in the intermediate sector applications (lighting, office equipment, drive motors for heating and air conditioning, and other appliances), it is intuitively appealing to consider direct current loads as a means of simplifying the PV system and reducing its cost. A study was conducted by the Research Triangle Institute of such an application under sponsorship of the U.S. Department of Energy as a part of its Photovoltaic Concentrator Applications Experiment Program (2). This paper describes only that portion of the work that dealt with the potential use of direct current loads and with the practical design consideration that governed the load choices.

2. Application Description

A new 44,000 ft^2 (4,089 m^2) engineering building consisting of offices and laboratories is being planned by the Research Triangle Insitute. Design peak loads for the building were: lighting--68 kW, services--60 kW, cooling--23.5 kW, miscellaneous--2.0 kW. The heating loads of the building were calculated to be quite small due to energy conservative design and contribution of the internal loads, particularly the laboratory equipment. The cooling system was designed to include an ice storage system so that on-peak cooling loads are negligible. An important objective of the design was to reduce the demand on the electric utility during its peak demand hours. For this reason, electric storage capacity was included to compensate for the periods of low insolation during peak demand periods that are experienced in the Research Triangle Park.

3. DC Load Components

The primary load components which could be implemented as direct current loads are the lighting and the cooling loads. The latter consists of electric motors which are readily available. Electronic ballasts suitable for fluorescent lighting common to commercial buildings are under development. These devices can be supplied by either alternating or direct current and are essentially small inverters that supply high frequency (5-35 kHz) power to fluorescent lamps. Because the conversion efficiency is higher than at 60 Hz, high frequency lighting has been under development for many years using 60 Hz as a source (3). Efficiency increases of about 15%, measured in lumens/watt, have been observed as shown in Table I.

TABLE I. FLUORESCENT LIGHTING EFFICIENCY (Lumens/watt)

Ballast	Lamp	
	Sylvania F 35	GE Wattmiser
G. E. Magnetic	62	65
Universal Magnetic	68	70
G. E. Maximiser	69	71
Iota 64 P*	76	79
Stevens**	79	84

* 120V High Frequency ** 277V High Frequency

Cost of the high frequency ballast was about $55 in 1978 compared to about $16 for a conventional 120V ballast (Luxor). However, mass production is expected to equalize the costs.

4. Photovoltaic Design Considerations

Two PV power systems were designed for the office/laboratory building, one for direct current (DC) and one alternating current (AC) distribution. These are diagrammed in Figure 1. Each system served the lighting load and cooling load from a 153 kWp array and includes 875 kWh of electric storage. Utility backup is incorporated in both systems. For the AC system, all loads are served through a commercial UPS inverter with a 280/120 VAC output. The DC/DC controller matches the array to the fluctuating battery voltage. Individual lighting circuits are 1536W each.

The DC system differs in several respects. The only available high frequency AC/DC ballasts are 120V input only. Therefore, the DC distribution system is 120V. A DC/DC controller is still required and a 120V/120V DC voltage regulator is necessary to assure constant illumination at varying loads. A DC/AC transfer switch is necessary to allow utility backup of the lighting DC loads, but no backup is provided for the cooling system since it has ice storage. The service loads of the building are served directly from the utility. Since conventional AC switches must be derated for DC, the lighting circuits are 960 watts each. Further, a larger number of motors are required for the cooling system because of the 120V DC restriction.

Careful analyses of the efficiencies of the various components and electric wiring losses showed that the AC and DC systems had virtually the same efficiency of 85% between array feeder and lighting ballast. Analyses of cost for the two systems is given in Table II based on catalog prices or manufacturers' estimates.

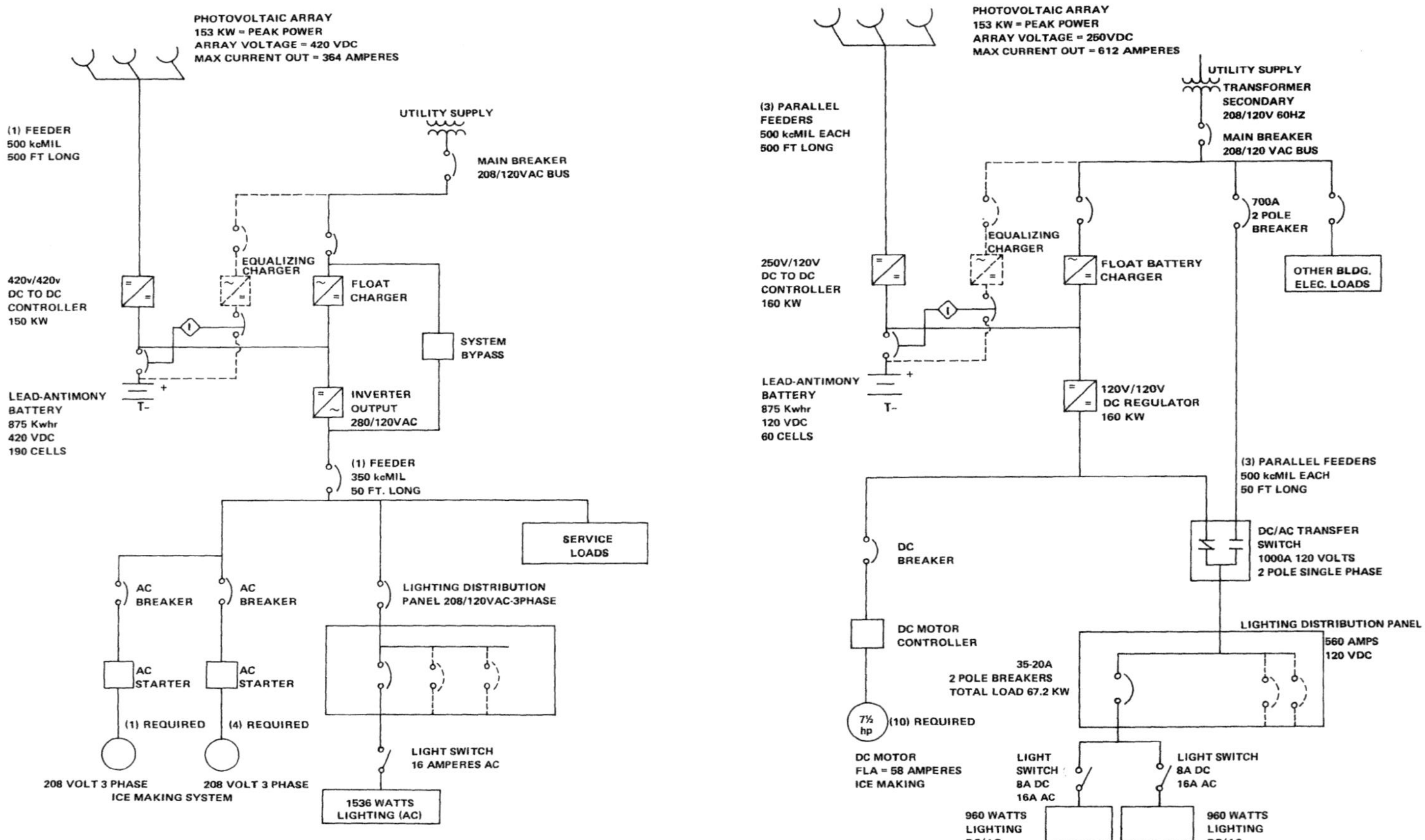

(a) AC DISTRIBUTION SYSTEM

(b) DC DISTRIBUTION SYSTEM

Figure 1. AC and DC System Systems.

TABLE II. COMPONENT AND SYSTEM COSTS FOR AC AND DC DISTRIBUTION SYSTEMS

Component	Distribution System AC	DC
Array Feeder	$ 5,350	$16,050
DC/AC Controller	30,000	30,000
Battery	145,000	145,000
Float Charger		20,000
Inverter	60,000	
DC Regulator		30,000
Distribution Feeder	500	1,605
Transfer Switch		12,000
Lighting Panel	1,290	2,580
Branch Wiring	7,500	8,878
Lighting Fixtures/Ballasts	61,465	61,465
DC Motor Controllers		8,000
DC Motors		30,000
AC Motor Starters	3,200	
AC Motors	4,950	
Total (less array)	$319,000	$365,578

5. Conclusions

It was concluded that the AC system was preferable since it offered the following advantages over the DC system:

- slightly lower system cost
- higher distribution voltage allowing smaller wiring, larger motor sizes and higher distribution efficiencies
- contactors and switches do not have to be derated
- AC motors require less maintenance
- equal total system efficiency
- utility backup is simpler
- compatible with standard building systems--better service, repair

The DC system does offer increased lighting efficiency (5-10%), and the ballasts can easily provide automatic or manual dimming, which can be a conservation feature.

REFERENCES

1. Applications Analysis and Photovoltaic System Conceptual Design for Service/Commercial/Institutional and Industrial Sectors, Report No. SAND 79-70-20, R. A. Whisnant, et al., September 1979.

2. RTI Photovoltaic Applications Experiment. Phase I--System Design. Final Report No. DOE/CS/94268-1, R. M. Burger, et al., March 1979.

3. "Progress in Static Converters for High Frequency Fluorescent Lighting," W. H. Johnson, Illuminating Engineering, June 1961.

4. Model 1A/D 120/120-2-40, Iota Engineering Inc., Tucson, AZ.

SESSION 2

US PILOT PROJECTS

Chairman : H. MACOMBER, Monegon Ltd.

- Lincoln Laboratory Experimental Photovoltaic Systems
- Terrestrial photovoltaic power system projects at Martin Marietta Corporation
- 20-Kilowatt El Paso photovoltaic project
- 400 X high efficiency solar module
- A 100 Kilowatt solar photovoltaic flat panel power system for the combined Berverly High School/C.H. Patten Vocational High School, Beverly, Massachusetts

LINCOLN LABORATORY EXPERIMENTAL PHOTOVOLTAIC SYSTEMS

DR. E. C. KERN, JR

Massachusetts Institute of Technology
Lincoln Laboratory
Lexington, Massachusetts 02173

Summary

Massachusetts Institute of Technology Lincoln Laboratory joined the U.S. Department of Energy (DOE) Photovoltaics Program in the fall of 1976. Since that time, the Laboratory has designed, built and operated four battery-equipped photovoltaic systems and two grid-connected systems (1.5 to 100 kilowatts peak). Currently, the Laboratory is managing U.S. DOE's residential system development activities and has nineteen grid-connected residential systems in various stages of design, development or evaluation. This paper provides a review of the characteristics of these systems.

BATTERY-EQUIPPED SYSTEMS

Battery-systems are chosen when the system is remote from conventional utility generation, when utility purchase of excess power is not advantageous, or when dc power is designated. Generally speaking, only the first criterion realistically imposes the need for on-site storage; pro-solar legislation and regulatory actions in the United States appear to mandate favorable economic treatment of grid-connected systems by electric utilities.

Battery-equipped systems characteristically comprise a photovoltaic array, a battery which can store several days worth of array output, and a dc-to-ac inverter which is sized to the maximum load imposed. Additionally, a system controller to protect the battery from overcharging and either to switch on a back-up power source or to disconnect the load to protect the battery from over-discharging must be incorporated.

The four battery-equipped systems currently being operated and evaluated by Lincoln Laboratory are listed below.

Northeast Systems Lat. 43°N (cont.)						
Solarex Prototype	14 4.6	8s x 10p	5.2	74.3	40°	6
TriSolar Corp Prototype	10.2 11.8	2p x 18s	4.8	47.6	45°	8
Westinghouse Prototype	16 2.3	12p x 13s	5.4	59.5	45°	6
Southwest Systems Lat. 33°N						
ARCO Solar Prototype	4.5 13	42s x 3p	6.6	80	25°	8
ARTU Prototype	16 2.3	7s x 6p x 2s	5.7	61.7	45°	-
BDM Prototype	15.8 2.3	13s x 9p	4.7	47.1	35°	6
General Electric Prototype	7.2 2.2	15p x 25s	6.1	73	26.5°	8
Solarex Prototype	14 4.6	8s x 11p	6.7	83.6	27°	6
TEA Prototype	15.8 2.3	14s x 8p	4.5	46.8	26°	6
TriSolar Corp Prototype	9 13	22s x 2p	4.5	58.2	30°	8
Westinghouse Prototype	16 2.3	12p x 13s	5.4	59.5	26.5°	6
John Long House	9 6	24s x 5p	6.6	71.2	20°	8
Southeast Systems Lat.						
Florida Solar Energy Center	16 2	14s x 12p	5	66	22.6°	2 @ 4
University of Texas at Arling-ton Prototype	16.5 .6	10s x 64p	4	107	27°	8
Hawaii Systems Lat. 21°N						
Pearl City	16 2.3	15s x 8p	4	45	30°	4
Molokai	16 2.3	15s x 8p	4	45	14°	4
Honolulu	16 2.3	15s x 4p	2	22	24°	2

BATTERY-EQUIPPED SYSTEMS

Application	Agriculture	Museum	Radio Station	Load Center
Location	Mead, Nebraska	Chicago, Illinois	Bryan, Ohio	Blanding, Utah
Latitude	41°N	42°N	42°N	38°N
Array Power (kW)	25	1.6	15	100
Array Size (m^2)	520	24	297	1642
Battery Capacity (kWh)	90	27	56	750
Back-up Power	Grid	Off-peak Grid	Grid	Diesel Generator

GRID-CONNECTED SYSTEMS

The U.S. DOE Photovoltaics Program is currently placing a high emphasis on grid-connected residential systems. The first of these systems begun operation at the University of Texas at Arlington in November 1978. Far simpler than battery-equipped systems, they rely on the presence of a stiff voltage reference signal to commutate the power inversion process. In such systems, the control requirement is simply to regulate the inverter operation as the array output varies from zero to maximum and as the utility voltage signal (amplitude, phase and harmonic content) varies slightly about a norm. Of special concern is the ability of such controllers to sense an open- or short-circuited utility line and to stop immediately the power inversion process.

Grid-connected systems have power converters which are sized to the maximum output of the array, not to the size of the load, thus system design is greatly simplified as the system is decoupled from the load.

RESIDENTIAL SYSTEMS

	Nominal Module $\left(\frac{V}{A}\right)$	Branch-Circuit	Nominal Array Power (kW)	Size (m^2)	Tilt Angle	Power Conditioner (kVA)
Northeast Systems Lat. 43°N						
System Test Facility	18.7 / 0.6	4p x 7s x 25p	7.3	130	45°	8
Lincoln Prototype	14 / 4.2	15s x 8p	6.9	89.2	45°	8
Lincoln House	14 / 4.2	14s x 9p	7.4	93.7	45°	8
GE Prototype	7.2 / 2.2	11p x 25s	5.8	73.3	33.7°	6

(con't)

TERRESTRIAL PHOTOVOLTAIC POWER SYSTEM PROJECTS AT MARTIN MARIETTA CORPORATION

M. S. IMAMURA
Technical Manager, Photovoltaic Systems
Martin Marietta Corporation

Summary

Three on-going major projects underway at Martin Marietta are the 470-kW Photovoltaic (PV) Power System for Saudi Arabia, 2.2 kW Actively Cooled Point Focusing PV Array, and Second Generation PV Concentrator. Previous work completed are Terrestrial Photovoltaic System Analysis for Tinker AFB, Oklahoma, PV System for Roxborough Park, 10 kW PV Concentrator Array Design, and Definition Study for PV Residential Prototype System. A brief summary of the current programs is presented.

1. INTRODUCTION

Martin Marietta's solar power experience during the past 12 years has evolved from an emphasis on space to that of terrestrial programs for DOE. Our photovoltaic (PV) system experience in the spacecraft includes testing of Mariner '69 sample panel, and SCATHA, Skylab, and classified satellites arrays. Various on-going and past programs on terrestrial PV systems are listed in Table I. The status of the current programs is summarized below.

2. 470-kW PVPS FOR SAUDI ARABIA

This system is the largest PV power system in the world that is being designed and fabricated for terrestrial application. System installation will start in early 1981, and it will be fully operational in mid 1981. The facility includes a 470-kW photovoltaic array, 1-MW Diesel-powered generators, 1470 kWh lead acid batteries, a 300-kVA inverter, and a solar and weather data monitoring station. The photovoltaic power system is capable of completely automatic operation. It is designed to operate in a stand-alone and cogeneration modes of operation.

The project started in January 1980. The contractual period is 47 months, which involves system design, fabrication, site preparation and installation, operation and maintenance, training, and system performance evaluation. To a large extent, design and fabrication will be accomplished at the Martin Marietta Corporation facility in Denver, Colorado.

Site preparation and installation will start in March 1981 and system checkout will be completed in July 1981.

3. SECOND GENERATION PV CONCENTRATOR

The objective of this program is to develop the current Martin Marietta Concentrator Array design into a more cost effective concept. Key design criteria are geometric concentration ratios of up to 100, use of silicon cells, and planar Fresnal lenses. Five prototype modules will be produced for test evaluation at both Martin Marietta and Sandia Laboratories.

A comprehensive computer analysis embodying all significant cell, lens and heat exchanger characteristics together with isolation and wind magnitudes and known imaging errors due to manufacturing tolerances has been developed. This, in conjunction with detailed cost data derived from our Saudi Arabia production program, optimizes lens size, "f" number and target size for minimum cost/watt. Design effort proceeded concurrently on a more cost effective cell laydown technique, individual heat sinks with integral interconnects and simplified mounting of the module to the support tube.

Using lens of 8.16" square an "f" number of 1.24 and concentration ratios of 84 a cost reduction of 37% \$/watt is predicted on the basis of 1000 arrays.

4. 2.2-kW ACTIVELY COOLED PV ARRAY

The actively cooled photovoltaic array is an evolution of the Martin Marietta passively cooled array which constitutes the principal power source for the SOLERAS 470-kW power system. It utilizes the same basic concentrating optical system comprising a parquet of 4 fresnel point focusing lenses operating at a power concentration ratio of 33:1.

A primary intent of this design is to produce useful thermal output in the form of hot water within a temperature range from 140°F to 190°F, in addition to the electrical output. With added insulation to the coolant flow passages, 50% of the incident solar energy is converted to useful heat energy resulting in considerably improved operating economies in instances where such thermal energy can be utilized. Approximately 11 kW of heat energy is provided per array.

The array is in final stages of assembly. Sun tracking controls are complete and have been satisfactorily tested in conjunction with the testing of individual photovoltaic modules. Processing techniques have been perfected to guarantee satisfactory cell/substrate/interconnect attachments with minimized series resistance. Resulting performance improvements compared to our original design amount to as much as 20%.

TABLE I LIST OF SOLAR PV PROGRAMS AT MARTIN MARIETTA

Title	Agency	Value	Period of Performance
SOLERAS PVPS, 470-kW	SERI	$16,500,000	01/80 to 09/83
Second Generation PV Concentrator	Sandia Laboratories	$ 353,300	03/80 to 03/81
2.2 kW Actively Cooled PV Concentrator Arrays	Sandia Laboratories	$ 261,200	09/79 to 08/80
Terrestrial PV System Analysis	WPAFB	$ 44,000	02/79 to 12/79
Roxborough Park PV Concentrators	DOE	$ 303,000	06/78 to 02/79
10-kW PV Concentrator	Sandia Labs	$ 414,000	07/78 to 04/79
Definition Study for PV Residential Prototype System	ERDA/ NASA LeRC	$ 92,000	11/75 to 12/76

20-KILOWATT EL PASO PHOTOVOLTAIC PROJECT

Dr. H. S. ZWIBEL

New Mexico Solar Energy Institute

SUMMARY

This paper describes the design, construction, and cost of the PRDA-38 20-Kilowatt El Paso Photovoltaic (PV) Project. The PV system is a fixed-tilt, flat-panel array that will shortly provide power to an uninterruptible power supply at El Paso Electric Company's Newman Power Station.

INTRODUCTION

This project is one of four PRDA-38 flat-plate experiments funded by the U.S. Department of Energy Energy (DOE).

The PV system, now in the construction phase, is sited at El Paso Electric Company's (EPE) Newman Power Station in El Paso, Texas, approximately one mile from the Texas-New Mexico border and 37 miles from the New Mexico State University campus at Las Cruces, New Mexico.

2. APPLICATION

The Newman Power Station's newest unit is a Westinghouse Pace 260 combined cycle plant, wherein a 110-megawatt steam turbine is powered from the waste heat produced by two 80-megawatt gas turbines. Computers are used to control the combined cycle facility; a Cyberex 25-kilowatt uninterruptible power supply (UPS) provides fail-safe power for these computers and various dc motors and relays. The PV system supplies power to the dc bus for the UPS.

3. SYSTEM DESCRIPTION

The array field is composed of six parallel-connected subarrays, four containing 11 panels connected in parallel and two containing 10 panels connected in parallel. Each panel is composed of nine series-connected Solar Power Corporation G361 modules; a module is 106 cm by 42 cm and consists of 36 series-connected cells. The output per module is 33 watts; panel output is 297 watts; subfield output is 3,267 watts for 11-panel subarrays and 2,970 watts for 10-panel subarrays; and total field output (at insolation of 1,000 watts/m^2 and cell temperature of 28°C) is 19.0 kilowatts. Simulated peak output from the field is 20.3 kilowatts.

The electrical output from the array is supplied directly to the load at 134 volts dc and a maximum current of 145 amperes. The load, approximately 165 amps, consumes the complete array output.

The photovoltaic array field covers a rectangular area approximately 30.8 m in the east-west direction by 15.8 m in the north-south direction. The configuration consists of three rows with 4 m spacing between each row. The rows are divided into separately switched and fused subarrays that contain either 10 or 11 panels.

The panel output for module voltage of 15.1 volts is 136 volts, which is the amount that must be produced at the array to supply the 134 volt dc load. Peak power is produced with module voltages close to 15 volts, which led to the choice of a panel composed of nine modules.

Pressure-treated pine is used for the panel structural material. It has these advantages: (1) is readily available and easily worked, (2) is ideal for small business fabrication, (3) has long life, (4) is a renewable solar energy resource, and (5) is relatively low in cost.

A simplified schematic of the system is shown in Figure 1. The annual energy supplied by the PV system operating at the voltage fixed by the bus is only 2.6 percent less than that produced using a perfect, loss-free peak power tracker. A life cycle analysis showed that a peak power tracker was not economically justified.

Voltage regulation is achieved without recourse to a voltage regulator because the array supplies less power than the load demands. The remainder of the load power demand is made up by the battery charger. Since the load's power consumption is always greater than the possible contribution of the photovoltaic array at the load's voltage, the array will be operated at the bus voltage.

The system controller, a GE Logitrol 550, cycles the array through its turn-on and turn-off cycles and detects PV system facility malfunctions. The system controller also monitors the array voltage and will close S_1 if the voltage rises too high. A reverse current detector

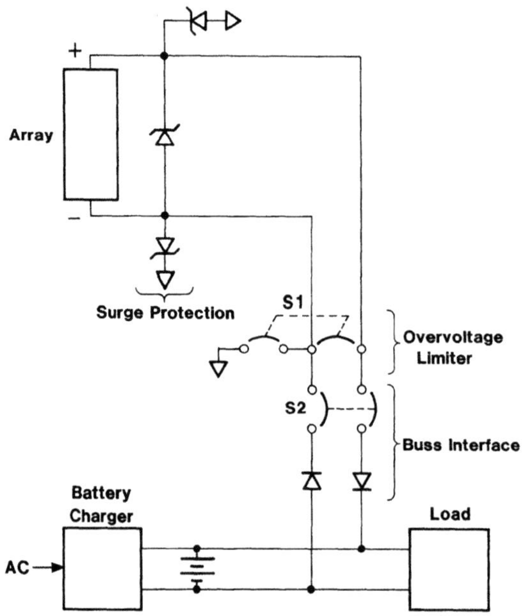

Figure 1. System Block Diagram

will also detect any current flow out of the UPS into the array and will cause S_2 to open if any reverse current is detected.

4. CONSTRUCTION EXPERIENCES

Fabrication and installation is proceeding according to design, schedule, and budget. Completion is planned for late October, with check-out requiring about one month.

The cost breakdown for the construction and installation estimate is given in Table I. The photovoltaic modules consume the lion's share of funds, namely $186,000 at $9.15/Wp (at 28°C, one sun) plus $0.22/Wp for shipping. The electrical system includes the cost of all wiring, switch gear, poles, conduits, etc., amounting to about $2.69/Wp.

5. FUTURE PLANS

In January 1981, the project will move into Phase III, at least two years of operation and evaluation. The extensive data acquired will be analyzed to determine system efficiency, degradation, agreement with computer simulations, fault detection and isolation, and reliability.

TABLE I

Cost Accounting Categories for Phase II

Cost Item	Cost ($)
Site	18,000
Collector/Array System	228,320
Electrical Systems	51,300
Instrumentation/Checkout/ Data Systems	37,538
Safety and Security Systems	3,000
TOTAL	$338,158

Acknowledgments

This project is being supported under DOE contract DE-AC04-79ET202629.

400 X HIGH EFFICIENCY SOLAR MODULE

Henri J.R. Maget
Varian Associates, Inc.
Palo Alto, California 94303

Summary

Photovoltaic systems generating electrical power and thermal energy can operate at high overall efficiency if the thermal energy is available at temperatures compatible with useful applications. High solar concentration, combined with GaAs solar cells operating at high temperature, offer such a possibility.

1. MODULE DESCRIPTION

The prototype module, shown in photograph 1, consists of twelve optical elements (plano Fresnel lenses) focused on twelve solar cells, with an average solar concentration ratio of about 400. The active area is 0.687 m^2 and the hexagonal configuration has been designed for maximum packing density. The module is liquid-cooled (water or water/glycol mixtures) and can be easily mounted, at four points, on a two-axis tracker.

2. MODULE ELEMENTS

Figure 1 illustrates the various elements, which include: 1) a Fresnel lens panel, produced from a constant-groove-depth, wide facet lens design, with transmission efficiencies up to about 85-87%, 2) a heat shield, to prevent damage of the plastic module parts during off-axis tracking, 3) packaged AlGaAs/GaAs solar cells, with individual efficiencies of 21-23% at 400 suns and at 50° C. The temperature effect on efficiency can be obtained from: $\eta = (0.22\text{-}0.24) - 0.035 \times 10^{-2}\, T_j$ where T_j represents the junction temperature, 4) a plastic housing, designed to allow module interlocking from two directions, for ease of maintenance, and 5) a coolant manifold. The thermal flux in non-uniform and emulates the photon flux distribution. To avoid local heating, an efficient jet impingement cooling subsystem has been designed (photo 2). The cell junction temperature is dependent on the boundary layer thermal gradient, ΔT_{BL} produced at the cell/coolant interface: T_j = T coolant + 12 + ΔT_{BL} and ΔT_{BL} can be evaluated from a semi-empirical equation, which predicts its dependence on coolant flow rate, G, and nozzle diameter, D,: $\Delta T_{BL} = f(G^2/D)^{-1.5}$

Under practical operating conditions, the thermal gradient between coolant and solar cell can be maintained at less than 20° C; the parasitic pumping power requirement is about 1-2% of the generated dc power.

The net module efficiency and generated power are illustrated in figure 2, for two different coolant temperatures, namely 28 and 50° C. Module efficiencies above 17% have been measured and shown to be sensitive to insolation rates.

For an insolation rate of 800 watts/m^2, the module produces 92 watts, and about 345 watts (equivalent) of thermal energy. The values estimated for a coolant fluid maintained at 100° C, would be 79 watts (e) and 360 watts (th).

3. APPLICATIONS

System designs, based on this type of module, have been conducted for three basic concepts: - electric- only - application of a GaAs photovoltaic solar system operating in parallel with a utility substation. This baseline system is typical of applications for dispersed power generation, supplying electrical energy only during periods of high demand. The coolant fluid is maintained at low temperatures and the thermal energy is rejected to the environment. Array field efficiencies of over 15% are possible - electric with thermal energy conversion - the use of a Rankine cycle heat engine, deriving thermal energy from the photovoltaic system, which operates at about 100° C, allows to increase the overall system efficiency to about 20%. - electric with maximum thermal energy recovery - application of a GaAs system for the desalination of brackish water, based on electrochemical membrane separation technology. In this case, the dc power operates a dc electrodialysis desalination unit, and the thermal energy produces ac power to operate auxiliary equipment, and is used to heat the brackish water, thereby reducing the specific energy demand, required for desalting water. The overall system represents the optimum recovery of solar energy.

Photo 1: Prototype Module Mounted on Two-axis Tracker .

Photo 2: Jet Impingement Coolant Component

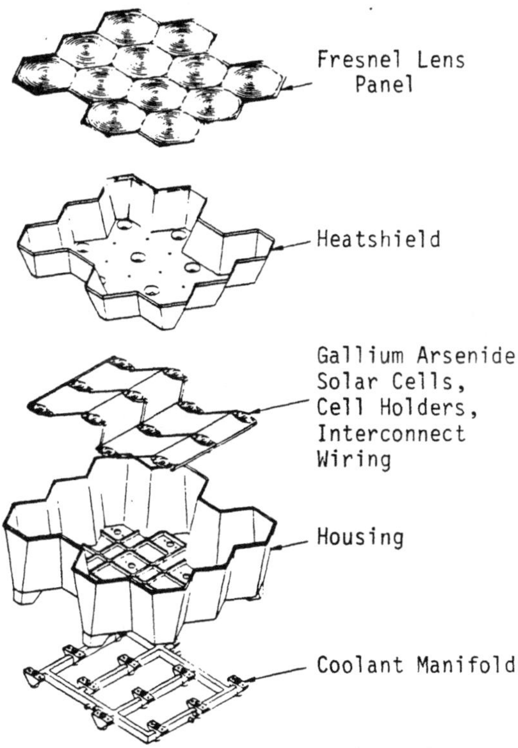

Figure 1: Solar Module Components

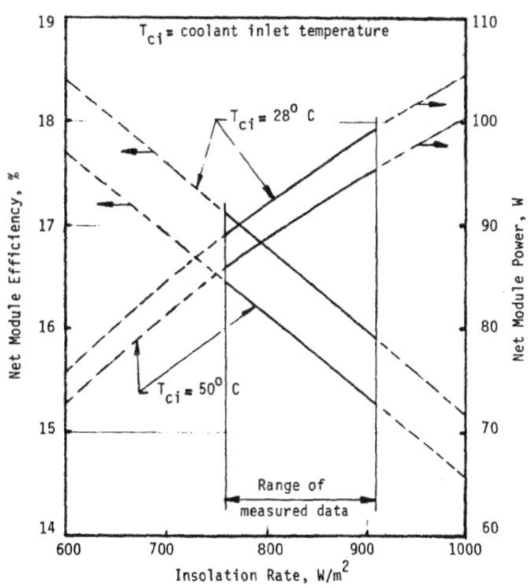

Figure 2: Net Module Efficiency and Generated Power

A 100 KILOWATT SOLAR PHOTOVOLTAIC FLAT PANEL POWER SYSTEM FOR THE COMBINED BEVERLY HIGH SCHOOL/C.H. PATTEN VOCATIONAL HIGH SCHOOL, BEVERLY, MASSACHUSETTS

R.R. ADDISS, JR.
Solar Power Corporation, 20 Cabot Road, Woburn, Massachusetts 01801

Summary

The 100 KW system consists of the photovoltaic array, the inverter/control subsystem, the building and utility interface, and the monitoring subsystem. The photovoltaic array consists of 40 separate subarrays of 80 photovoltaic modules each (5 parallel by 16 series), deployed in rows on the southerly facing slope north of the school building. Typical operating voltage is 250 V. The wiring scheme permits individual modules to be disconnected without a radical change in subarray output current. Power is transmitted at 4160 V from the inverters and a step-up transformer to the main 4160 V utility feed line in the school. Separate metering measures power bought and sold. At the optimum tilt angle of 40°, the array provides 116 MWh of AC energy annually, or 8% of the school load. It provides a good temporal match to the summer daily load profile, resulting primarily from the central air-conditioning. Some power is sold to the utility during the summer, mainly on weekends. The immediate impact is a $4000 saving in the annual utility bill. A fault detection and isolation scheme which can find faulty modules is incorporated into the monitoring subsystem. Safety during installation, operation and maintenance was a major consideration in system design.

1. INTRODUCTION

This system was designed and is being installed under the sponsorship of the United States Department of Energy as part of its PRDA 38 program of system application experiments. Funding of the present system construction phase is provided under contract No. DE-AC04-79ET20631. Solar Power Corporation is the prime contractor to DOE on this project. Other participants are Stone & Webster Engineering Corporation of Boston, the New England Electric System of Westborough (the cognizant electric utility), and the Beverly Public School System.

2. APPLICATION

The Beverly High School is typical of a high school in a suburban community. The City of Beverly is a coastal community 18 miles north of Boston in the State of Massachusetts. About 2000 students attend school in this 280,000 square foot, multiunit building sited on a 22 acre plot of land. The curriculum includes a vocational school program which utilizes about 50,000 square feet of the building. Adult education courses are offered in the evening, and an 8 week summer program provides a year-round electrical load. The main summer load arises from the central air conditioning of one 25,000 square foot wing.

The grounds are used mostly for athletics, but about three acres of an unused, south-sloping hill just north of the school building will be used to deploy the photovoltaic array. Original plans to site the array on the roof of the school building were abandoned after a structural analysis showed that major, costly changes in the roof and its supports would be required to accommodate the added wind and snow loads.

3. SYSTEM CONFIGURATION

a. Photovoltaic Array - The fundamental building block of the array is the Model G12-361-CT photovoltaic module manufactured by Solar Power Corporation, containing 36 series-connected 100 mm diameter silicon solar cells. This is a tempered glass superstrate module with a silicone rubber encapsulant. Two bypass diodes are built into each module, each bypassing half of the cells. This limits cell temperature due to reverse bias power dissipation to a maximum of about 100°C, which is well below the survival temperature limits of this module.

Each subarray contains 80 modules, formed by first wiring a group of 5 modules in parallel and then connecting 16 of these groups in series. All interconnect wiring is located within subarray junction boxes, and individual modules have environmental connectors which plug into the junction boxes. An additional bypass diode across each parallel group of 5 modules is located within the junction boxes. Individual modules can therefore be unplugged from the junction boxes without open-circuiting the subarray, eliminating arcing at the connector and only slightly reducing the power output of the subarray. Each subarray also has its own disconnect switch which can remove the subarray from service either by open-circuiting or short-circuiting the subarray, without affecting the performance of other subarrays. In the "open" position, with all modules unplugged, work can safely be done inside the junction boxes. In the "short" position, no voltages higher than 15 volts can arise at any point in the subarray (n - 1 diode drops for the n = 16 bypass diodes in series within the junction boxes), so the modules can be safely washed or handled. Meter shunts in series with the negative ends of each subarray provide a signal to monitor the subarray current, providing a means for fault detection and isolation. All subarrays (negative end) are grounded at the inverter input through a ground-fault detection circuit which shorts the DC bus at the inverter input if the ground current exceeds a preset level. Care has been taken in system design to minimize naturally occurring ground currents so that the ground-fault interruptor can be set at a sufficiently low level to protect personnel from accidental contact with high voltages.

The entire array consists of 40 separate subarrays which are deployed in rows on the hillside. These are divided into two separate subfields of 20 subarrays each. Each subfield has its own inverter and control system, and therefore the two subfields operate completely independently of each other. Only the AC outputs of the two inverters are interconnected (in parallel). At 100 mW/cm^2 insolation, each subarray has a peak power rating of 10 A at 250 V, which results in the 100 kW rating for the entire array field.

b. Inverter/Control Subsystem - The 60 kW inverters manufactured by Delta Electronic Controls Corporation were chosen for this application. These units include all of the controls and alarms required for complete operation of the system, and produce excellent quality power. System operation is completely automatic, including startup at daylight and shutdown at night, and shutdown upon a fault occurrence (including a utility outage).

Automatic operation includes peak power tracking of the array output. The control system can also be operated in a manual mode to facilitate system diagnostic and maintenance functions. Alarms to indicate fault conditions are also included. The DC input to each inverter is the output of 20 subarrays in parallel. The AC output is at 480 V, 3 phase.

c. Building and Utility Interface - The inverters are housed in a shed at the bottom of the hillside, near the front edge of the array field. A transformer at this location converts the inverter outputs to 4160 V for transmission to the school and interconnection directly to the main 4160 V utility feeder line. Rachet type, pulse counting watthour meters installed in this feeder line measure energy exchanged (bought and sold), and enable the utility to carry out its purchase and sale agreement. In addition to the special metering, an automatic disconnect switch which can only be closed by the presence of utility power will be installed. This will absolutely assure utility personnel that the school cannot energize a "dead" line.

d. Monitoring Subsystem - The heart of the monitoring system is the computerized On-Site Data Acquisition System (ODAS). This will be the primary means for obtaining the technical data needed for the system performance evaluations. It also provides a means for system diagnostics. Input data include the 40 individual subarray currents, school energy demand and net energy flow, power conditioning performance, meteorological data from the meteorological station erected within the array field, solar cell operating temperature, and system status. The ODAS will also perform the fault isolation function by comparing subarray currents and outputting a location message if any are below the expected level. Individual subarray I-V curves may be recorded by sweeping the inverter input voltage by means of its manual override control.

4. SYSTEM PERFORMANCE

The expected performance of the system was calculated using the weather and insolation data on the SOLMET computer tape for Boston. On this tape, many years of data have been reduced to a "typical meteorological year". The known efficiencies of various system components (e.g., the modules and inverters), including their variation with temperature and insolation, as well as wiring and degradation losses were taken into account. Results were then correlated with the recorded building load profiles.

At the optimum fixed tilt of 40°, the system produces 116 MWh of AC energy annually. This is about 8% of the total school demand. 96% of this output is consumed by the school directly, and the remaining 4% is sold to the utility. Almost all of this excess energy is produced on summer weekends. The weekday summer load, due primarily to the central air conditioning, provides a good temporal match to the array output. The array output and building load comparisons are summarized by season in Table 1.

TABLE 1. ARRAY OUTPUT/BUILDING LOAD COMPARISONS

	Summer	Fall/Spring	Winter	Annual
Building Load (MWh/Day)	1.9	4.1	4.3	3.8
Array Output (% of Load)	20	8	5	8
Distribution of Array Output				
% To Load	93	100	100	96
% To Utility	7	0	0	4

SESSION 3

EC PILOT PROJECTS

Chairmen : G. BEER, Phoebus, Italy
Y. CHEVALIER, Commissariat à l'Energie Solaire, Paris, France
R. KOEPKE, Kernforschungsanlage Jülich, Germany
W. SCHNELL, Commission of the European Communities, Brussels, Belgium

OVERVIEW

- European Communities Photovoltaics Pilot Plant Programme

PROJECT PROPOSALS

- Multi-purpose high efficiency concentration photovoltaic systems
- ⁑ Photovoltaic power plant in Denmark
- ⁑ Photovoltaic power pilot plant in the Island of Kythnos - Greece
- * Realization of a 5 kW photovoltaic generator
- ⁑ 300 kW Photovoltaic pilot plant Pellworm, Germany
- * Sophia Antipolis 5 kW photovoltaic power generator for habitat
- A 50 kW photovoltaic power plant coupled to the 15000 V electricity grid
- ⁑ Photovoltaic project for Giglio Island
- ⁑ Conversion and storage of solar energy by means of hydrogen
- ⁑ Renewal of rural life by photovoltaics, a photovoltaic pilot plant for rural communities
- ⁑ A 60 kW photovoltaic rural electrification for the Island of La Réunion
- * The 5 kW generator at the Alpine refuge "Les Evettes"
- Power supply for a solar water heating installation

xx - Photovoltaic pilot plant for the water supply of the Tremiti Islands

- The Epsilon project, European photovoltaic system installation in London

xx - Self-supporting photovoltaic supply system for the school of maritime studies on the Island Terschelling, Holland

xx - Adrano photovoltaic project

xx - Alicudi project

xx - Scolar project - The study of a photovoltaic system to identify the potential for solar electric generation in the UK

xx - Power supply of electrical energy control of an airport

xx - Power supply for FM and TV emitter station

xx - The Fotavoltaic project - A 50 kW photovoltaic pilot plant proposed for Fota Island, Cork

xx - Photovoltaic power plant Zambelli (Verona)

xx - North Europe hydro-photovoltaic peak power producing plant

xx - Photovoltaic utility connected to the grid, feeding the auxiliaries of a solar thermal system

xx - Aghia Roumeli - Island of Creta - A 50 kW autonomous photovoltaic power

- Solar photovoltaic cold store

- Photovoltaic unit integrated with a hydrogen generator, pressure storage and distribution system

- Photovoltaic unit integrated with an active chlorine generator, storage and distribution system

- A 30 kW photovoltaic power generating plant for industrial application

- 80 kWp solar photovoltaic plant for isolated areas and for integration in multisources energetic economies - designed by COCSIER Int. Consortium, Italy

x *These projects have already been constructed as part of the Commission's first solar energy R&D programme (1975-79)*

xx *On 26 November 1980, The Commission selected these projects for its pilot plant programme (1980-83). The technical features of some of them have been slightly changed since their presentation at the workshop in Sophia Antipolis.*

EUROPEAN COMMUNITIES PHOTOVOLTAICS PILOT PLANT PROGRAMME

W. PALZ
Commission of the European Communities
Brussels, Belgium

Photovoltaic pilot projects in the power range around 100 kW is a completely new subject in Europe. When the Commission first announced its intention to support projects in this field, it encountered a tremendous interest from industry and also from the national authorities of most of the EC member countries.

The purpose of this programme is system R&D which is the necessary complement of the European Communities' programme on photovoltaic components which has been going on since 1975. The programme of these EC photovoltaic pilot projects has been outlined since 1977. It was finally decided in August 1979 by the Council of Ministers as part of the European Communities' second 4 year energy R&D programme. Initially, the programme included the construction of a 1 MW photovoltaic power plant in one block, in addition to the distributed projects whose power totals up to 1 MW as well, and which are going to be implemented now. The 1 MW power plant in one block had to be dropped as a result of budgetary restraints in the Council's decision. In February this year, the Commission published a call for tenders on the pilot projects and received 31 proposals with a total power of 2.8 MW and a total cost of 45 million EUA*. All projects which will have been selected for construction will be completed by the middle of 1983 at the time when the current second EC solar energy programme ends.

In association with this call for tenders, the Commission has undertaken an investigation among solar panel manufacturers in the European Communities. More than 10 companies have been identified. They project for the year 1982 a total production capacity of about 5 MW for solar cells and panels. Both monocrystalline and polycrystalline silicon cells will be employed. A five year guarantee is generally given for panels. Price quotations for 1982 range from 10 EUA per Watt to about 7 EUA per Watt depending on companies.

* 63 million dollars

The power range for the submitted pilot projects is between 30 and 300 kW. Proposals have been received for all climates and countries in the European Communities including Greece, which will join the European Communities on 1st January next year.

Concentration has been given less priority than flat-plate arrays due to weather which in many regions in Europe is frequently overcast. It can be expected that only one out of all selected projects will include concentration, most probably troughs. The cost per peak Watt in most proposals was approximately 20 EUA; that is including R&D, manufacturing, and installation of the complete system.

The Commission is ready to support the projects which will be finally selected with a maximum of 7 EUA and possibly a bonus. The rest of the cost will be provided by national governments, regional authorities, industry, the users of the power plants and in many cases, national or local utilities. It is very encouraging that the utilities of most countries have shown interest in these projects and will be involved in their construction.

Projects will be implemented by international consortia made up from companies based in the member countries of the European Communities. As stated earlier, focus in the programme will be on R&D for systems and sub-systems. Sub-systems include power conditioning, structures, interfaces with a grid or the users, etc. For most proposed projects, batteries have been envisaged for storage with an average capacity of one or two days.

It has been discovered that hydrogen storage systems in association with photovoltaic power plants seem not yet to be ready for use in the short term as only 3 projects involving hydrogen have been proposed at a rather high cost and a general hesitation to submit hydrogen projects could be felt. In no proposal has a fuel cell been put forward.

To ease the storage problem, many projects foresee the combination with wind generators, hydro-electrical plants, diesel generators or the grid.

The Commission feels that it is very important for applications to be sound and for the plants to fit nicely into the environment. As asked for by the Commission, a wide variety of applications has been proposed, such as :

- electrification of islands or small villages;
- radio emission, T.V. studio;
- airport management;
- water desalination on islands;
- ice-making, cooling;
- dairy;
- water pumping for urban use;
- heat pumps;
- hydrogen production for industrial use.

Many of the projects will not be connected to a grid.

It is expected that the final decision on these projects will be taken towards the end of this year. Probably more than a total of 12 projects will be selected for construction.

MULTI-PURPOSE HIGH EFFICIENCY CONCENTRATION PHOTOVOLTAIC SYSTEMS

G. BERTELS

E. N. I. (Electro Navale et Industrielle N.V.) ANTWERP - BELGIUM

1. INTRODUCTION

Based on the experience we gained in thermodynamic solar systems with concentration by means of parabolic-trough thin glass mirrors, our company started with development of photovoltaic cells for concentrated light (50 suns), to produce directly the required electrical energy.

The concentrating solution may be considered, from the economical point of view, only in case of using high efficiency cells along with a low priced system.

We therefore chose to develop an IBC- (Interdigitated back contact) cell, ion-implanted on a single crystal silicon, which can theoretically reach efficiencies of about 24 % under high concentration. Moreover, the ion-implantation allows high automation. This type of cell has a very low silicon cost, owing to the concentration ratio.

All those characteristics allow to produce a low cost system.

On the other hand, a thermal recuperation of the heat losses can easily be achieved in the cooling system of the cells.

2. SOLAR CELLS

The improved IBC-cell, especially designed by E.N.I. for use under high concentration ratio, offers many advantages versus conventional cells :

- The serial resistance is lowered, which is very important under concentration. The electrons are collected exactly where generated.
- The open circuit voltage Voc is not saturated.
- The front surface is completely available for light conversion.

The minority carrier collection is for all these reasons much greater, and the conversion efficiency is strongly improved, with a very good fill factor.

Optimization of some parameters, like the fingers length, the lifetime, the cell thickness and the low surface recombination, will again improve the efficiency of the cell.

The main characteristics of the E.N.I.-cell are as follows :

- The substrate is N-type with supplementary p-junction on the illuminated front surface, which gives a tandem cell. Note that this P+N structure is better than the N+P, because of the higher Voc under high injection.
- The crystal orientation is < 100 > for front surface chemical texturization and etching.
- The thickness is about 100 μm, obtained by chemical etching of a 330 μm-wafer. The obtained thickness becomes then smaller than the carrier diffusion length.
- The carrier lifetime is about 1000 μs before treatment for the 250 Ohm-cm-wafers and about 200 μs for the 10 Ohm-cm-wafers.
- The ion-implantation is chosen for the precise control in junction creation and for the possible automation.
- The metallization is realized by means of Ti-Ag evaporation; followed by Ag-electro-plating.
- The combined AR (anti-reflective)-coating and texturization improves the efficiency by about 30 % with respect to an untreated cell.

The analyses of the various resulting parameters after fabrication in the research laboratories of the Catholic University of Louvain-la-Neuve (U.C.L.) is not yet completed. A new Research and Development-Program is to be started for optimization with Belgian State R & D funding program.

Nevertheless the results are already strongly encouraging. Efficiencies of about 18 % under AM 1 by 25°C are measured in laboratory with a Voc of 575 mV. Peak efficiencies of about 19,5 % with a Voc of 630 mV, are registrated with CR = 8 but without incapsulation or cooling. Tests will be performed next time under a concentrated light of 50 suns.

3. CONCENTRATOR SYSTEM

A parabolic-trough concentrating mirror is used to get the concentration ratio of about 50. The whole system is constituted by 11 mirrors of about 55 cm width and 2 m long, which are assembled side by side on a common single pedestal, equiped with a 2-axis tracking system.

The thin glass mirrors (Glaverbel made) are glued on a composite sandwich of galvanised steelplate and aluminium honeycomb by means of an epoxy paste.

The parabolic-trough shape is obtained under combined vacuum and steam pressure in a moulding machine, adapted to an industrial mass production.

The optical characteristics of the mirrors are checked with a laser spot before installation.

Such a sandwich structure has a high mechanical resistance; a violent local impact (for instance hammer, stones ...) involves no cracking propagation and assures maintained mirror optical properties, all around the impact point.

The cells are positioned on the focusing line of the concentrators; they are incapsulated in a groove, on the lower face of an extruded aluminium hollow profile, which acts as support and cooler. Incapsulation is performed by means of a thermally conductive epoxy, a transparent silicone gelcoat, and a thin glass.

A set of 20 serial-connected cells reach a nominal voltage of about 12 Volt. Further external serial-parallel connections are made in order to obtain the required voltage.

The cavity of the support is filled with cooling fluid, like FREON 113 or ethylic alcohol. Their boiling temperatures are constant and will maintain a constant temperature at the back side of the cells, along the 2m-length collector.

The produced vapour condenses in the heat exchanger surrounding the upper side of the supporting tube. The condensed vapour flows back down after water cooling.

The power delivered by a module, like the one we present on scale 1/10th in the stand of the European Commission in Cannes, is about 1650 W peak.

The complete system we propose to install will have a minimum rated power of about 50 kW peak.

4. SYSTEM AND APPLICATION

The first application envisaged by our company for this demonstration project is to supply solar electrical energy to an hospital, connected with the lighting grid, or for special application like emergency power. Lead-acid batteries and DC/DC convertor are included.

The connection with the whole grid is not absolutely necessary. The warm water could be stored in a boiler.

A small water electrolysis installation could be added for hydrogen, and oxygen production. However this is not included in the project.

We propose to install this 50 kW system near the University Hospital of the Catholic University of Louvain in the Brussels suburbs. The University is ready to place the ground at our disposal and agrees to pay for both the produced electrical energy and the warm water, on a market price basis.

On the other side, an implantation in a Southern Europe site, where direct illumination is much larger than in Belgium, could also provide an interesting demonstration.

On such a site, the energy balance during the year between flat panels and concentration devices would lead to very important data.

The yearly available energy in kWh for grid connection or emergency power, will indeed decide of the interest of a photovoltaic device for such applications in the future.

We think that our " Solar - E.N.I. 1 " system, owing to the high efficiency of the E.N.I.-cell, could deliver much more kWh per year than normal flat panels of equivalent area.

Moreover it will recuperate an important amount of calories in warm water and demonstrate a very high level of global hybrid efficiency.

Photovoltaic Power Plant in Denmark

ERIK GREBY, GERHARD BRUNING and HARTWIG WILLMES
Danish Energy Technique Ltd., SIEMENS Ltd. and
VARTA Batterie Ltd.

Introduction

In the following is given a general description of a 100 kW photovoltaic power plant contemplated to be constructed in Denmark.

Companies and Organisations behind the project

The companies proposing the project are:
SIEMENS AG, VARTA BATTERIE AG and DANISH ENERGY TECHNIQUE Ltd. while the 2 Danish public utilities ELSAM and ELKRAFT are sponsoring the project together with the Danish Ministry of Energy. The board of directors of the utilities announced their sponsorship in June and August 1980, and the Ministry of Energy issued a letter of intent of September 23rd, stating the Ministry's interest in the project, all provided the project will be supported by the Commission of the European Communities and by the Bundesministerium für Forschung und Technologie.

In order to secure Danish participation on the highest technological level, the Danish Technical University and "Elektronikcentralen", which is an institute under the Academy of Technical Science, have been involved alongside with Danish Energy Technique Ltd. in the preparatory work up to now, and will in the future participate in the establishment and survey of the plant.

The Danish industry will be involved to the greatest possible extent with a view not only to the evident immediate interest in job-opportunities, but also - perhaps mainly - with a view to future export possibilities.

Finally it has been proposed that Elkraft, which is the utility responsible for the supply of electricity on the Island of Zealand including the capital of Copenhagen, shall run the plant during a 5 - 10 years test-period. Elkraft has conceded to this proposal.

Location

In order to select the most appropriate location for a possible photovoltaic power plant with a peak-effect of 100 kW, a series of criteria was listed:

- The plant should substitute an existing transformer station between 75 kVA and 125 kVA.
- The transformer station should be without ring connection to the network in order to simulate Island conditions.
- Yearly consumption should be approx. 150.000 kWh.
- For meteorological reasons the location preferably should be no more than 1 km from the sea and north of the shore.
- Average yearly sun-hours to be maximum.

These criteria were used as input in an EDP-programme, runned by Elkraft, and eventually lead to the selection of a village Vester Bøgebjerg, some 100 km south-west of Copenhagen, approx. 1 km from what proudly is called the Danish South-seas.

Environmental authorities presently are scrutinizing the proposed location, which as to verbal information from the Ministry will be met with approval.

Purpose of plant

We propose the installation of a photovoltaic system with a peak rating of 100 kW in Denmark. The aim is to prove that solar energy can be utilized as a reliable source of electrical energy for small isolated consumers.

Design of the plant

A solar generator (SG) consisting of approx. 1200 m^2 solar panels with a rating of 100 kW at AM 1 and 25 ^oC forms the core of the pilot project. The expected efficiency will be in the range of 12-14 %.

The output of the solar generator fluctuates during the day and is also dependent on the season and the weather. For this reason, a storage battery is necessary to compensate for these fluctuations.

We use led-acid batteries specially developed for photovoltaic systems. These batteries have a service life of more than 10 years, require little maintenance and are characterized by a low self-discharge rate and high cycling ability.

These features can be reached by using alloys of very low antimony content combined with a gas recombinator system.

The solar generator is connected to the d.c. bus via a power matching unit (PMU). The battery is also connected to the d.c. bus. The function of the PMU is to ensure that the solar generator always operates at its maximum power point and, in case of an excess of available power, to protect the battery from overcharging.

For economic reasons, the capacity of the energy store is limited. The battery in our pilot project would have a capacity of 1000 kWh and could supply the consumers for approx. 30 to 50 hours.

The state of charge condition of the battery will continuously be monitored by a microprocessor control unit.

Normally a diesel generator would supply the system during long bad weather periods. In our proposal we shall utilize the available power system.

An inverter (PSU) is used to provide a three-phase, 380 V, 50 Hz supply for the consumers.

Special circuit arrangements for the power converters (PMU and PSU) are under development which will allow the operation at an efficiency as high as possible particularly in the partial load range. According to latest calculations the expected efficiency will be higher than 96 % over the whole load range which is much higher than nowadays standard equipment efficiencies as a result of intensive research and development.

All subsystems of the photovoltaic pilot plant will be monitored and controlled by a superordinated centralized control system, which always ensures optimum economical operation.

The electronic equipment as well as the battery are assembled in specially designed housings of container type to facilitate transportation, make the installation easier and allow maintenance and protection of the parts.

Photovoltaic Power Pilot Plant in the Island of KYTHNOS - GREECE

G. Bruning, C.N. Kourogenis, H. Willmes

Introduction

A 100 kW photovoltaic power generation unit is proposed for the Island of Kythnos to operate in parallel with the existing diesel grid and a wind energy system of 60 kW nominal power.

The aim of this plant is to prove the technical and economic feasibility of parallel operation of above system, and to optimize the quality and quantity of energy so generated.

Organizations and Companies involved in this project

The organizations are:
The Scientific Research and Technology Agency (YEET), The Hellenic Industrial Development Bank and The National Energy Council of Greece.

The companies are:
SIEMENS AG, The Puplic Power Corporation of Greece and VARTA BATTERY AG.

The Bundesministerium für Forschung und Technologie (BMFT) is expected to participate in this project.

Local firms will be involved in this project as far as possible under the supervision of PPC or Siemens Hellas.

The Island of Kythnos and the Present Power Situation

The island is located 65 kilometres south of Athens and is one of the Cyclades islands. The population is about 1800 inhabitants dealing with animal breeding, fishing and tourism.

The installed capacity of the island's grid consists of five diesel sets with a total rated output of approximately 500 kW. This capacity is expected to be raised by 60 kW next year with the installation of a separate pilot plant utilizing wind energy under a bilateral contract between M.A.N. and P.P.C.

The annual consumption of electricity is about 950.000 kWh and the proposed photovoltaic plant will deliver approximately 170.000 kWh per year which would mean that under the proposed mode of operation more than 60.000 litres of fuel oil could be saved per annum.

It is expected upon completion of the project that the renewable forms of energy (photovoltaic and wind) will cover the electricity needs of the island for long periods of time substituting at least three of the above diesel units.

This island was selected for the photovoltaic pilot plant because it is situated within the geographical belt which promises the best efficiency for solar energy conversion (average solar radiation 4.65 kWh per m^2 and day) and there are more possibilities of large scale photovoltaic applications not only in the vicinity (Mediterranean sea) but also in other similar places of the world which are favored with high solar radiation.

Technical data and mode of operation

The solar photovoltaic generator consists of about 50 units with a total surface (active solar cell surface) of 1200 square meters. Its rating is 100 kW at AM 1 and 25°C.

The panels will be embedded in standard aluminium frames which will be produced in Greece. They will form arrays fastened on concrete foundations and will meet all the special requirements of environmental conditions.

The electrical energy produced by the panels is fed to the grid via a specially designed modular inverter and/or via a special D.C./D.C. converter to the storage battery depending on the energy production - consumption pattern on the island. It will be controlled in such a way that the optimum economics of the overall operation of the combined system is achieved.

Further advantageous fact is that the diesel generators can be kept in operation at their highest efficiency level covering the basic load, whereas the solar unit in combination with the battery will cover all abrupt variations in the load.

The components of the system

The solar panels are now under development and they will include a new flat panel technology using silicon as base material. The expected efficiency will be in the range of 12-14 % at AM1 and 25°C.

The DC/AC and DC/DC converter are of a modular type. The first one satisfies the primary operation to feed the energy into the grid, the other converter adjusts the voltage between the DC line and the battery and feeds the storage subsystem with surplus energy or assists in a controlled discharge of the battery when additional energy is needed. Both converters will interrupt the DC line when neither the grid can consume nor the battery can store the energy produced by the solar cells.

Special circuit arrangements for the power converters are under development which will allow operation at an efficiency as high as possible particularly in the partial load range.

According to latest calculations the expected efficiency as a result of intensive research and development will be higher than 96 % over the whole load range which is much higher than standard equipment efficiencies today.

Both converters will satisfy the condition that maximum power is drawn always from the panels by means of a maximum power point control unit.

The lead acid battery will have a total capacity of 600 kWh and will be specially developed for solar plant purposes.

The main characteristics are

- more than ten years of service life
- ability of cycling
- low self discharge rate
- low maintenance

These features can be reached by using alloys of very low antimony content combined with a gas recombinator system.

The state of charge condition of the battery will continuously be monitored by a microprocessor control unit.

All subsystems of the photovoltaic pilot plant will be monitored and controlled by a centralised control system (C.C.) which ensures always the optimum economical operation of the whole power system.

The electronic equipment as well as the battery are assembled in specially designed housings of container type to facilitate transportation, make the installation easier and allow maintenance and protection of the parts.

REALIZATION OF A 5 KW PHOTOVOLTAIC GENERATOR

D. Keaveny
AEG-TELEFUNKEN
Fachbereich Neue Technologien, Raumfahrt
Wedel / Holstein

Under the auspices of the Commission first Energy R + D programme and the Germany Ministry of Research and Technology a 5 kWp solar generator was designed and installed in Berlin in 1979.

The generator consists of 486 modules composed of square shaped solar cells. These 10,3 Wp modules are electrically connected 243 in parallel and 2 in series. They have a peak output power at AM1 of 5000 Wp and a nominal voltage 24 V DC. The total cell area is approx. 44 m².

The modules are mounted in groups of 14 on lightweight aluminium frames, which in turn are mounted on galvanized steel girder superstructure. The frames are set at 60° angle of tilt and are designed to bear up to wind speeds of 200 kmh. They are mounted in 3 rows and in terraced formation to avoid shadowing (see fig. I).

A lead acid battery consisting of 12 stationary type cells each of 2 V 1000 Ah (24 kWh) acts as storage and reserve for bad weather periods.

The battery is installed in an especially constructed cabinet with an automatic controlled gas extractor. This cabinet together with the measurring equipment and the broadcast transmitter are located in a container beside the solar arrays.

The broadcast transmitter has a continuous input power requirement of 600 W 24 V. This load profile is compatible to the mean solar generator output allowing for approximately 5 hours operating time in January and 24 hours in June.

Fig. II

The following data are being continuously measured and recorded:

- Output power of the solar generator in Wh
- Battery charging current Ah
- Load consumption AH
- Battery discharging current Ah

Single modules are also being plotted

- Open circuit voltage.
- Short circuit current.
- Cell temperature.
- Ambient temperature.

Further to these environmental influence on the whole system.

- Industrial dust and dirt on the power output.
- Deterioration through climatic condition and service life of all parts.
- Ascertainment of electrical losses.

In June 1980 the solar generator was modified for high voltage operation. The modules have been connected so that an open circuit voltage of 400 volts is obtainable. This was necessary to run tests with a newly developed inverter which operates without a battery and is capable of tracking the MPP of the solar generator. First test indicates a very high efficiency of more than 90 % in part or full load conditions.

300 KW PHOTOVOLTAIC PILOT PLANT PELLWORM / GERMANY

V. Cordes and K. H. Korupp
AEG-TELEFUNKEN
Fachbereich Neue Technologien, Raumfahrt

Summary

The 300 kW solar power system described in this paper shows a feasible and suitable application for solar systems in our latitudes. The fore-seen consumer strongly influenced through tourism during the season has a load profile which is very corresponding with the insolation profile during the most time of the year. It will be a self-supported solar powered system with the utility grid as power back-up. The loc-ation offers the possibility of a parallel connection with 9 different wind generators, which are already installed and are presently being tested. A certain amount of solar generator modules will be mounted on a two-axis tracking device in order to investigate the utility of tracking systems in our latitudes.

1. System Application

Pellworm Island is one of the North Sea holiday Islands belonging to Schleswig-Holstein/Germany. In order to demonstrate the usefulness of solar power systems on tourist islands like Pellworm a model consumer with mixed loads and a load profile which is well corresponding with the insolation profile has been chosen, i.e. the recreation centre of Pellworm. The recreation centre is consisting of restaurant, indoor-swimming-pool, sauna with massage department and community rooms. The daily load profile has two daily peaks in the summer, one occurs around noon and the other in the later afternoon. In the wintertime there is only one peak in the late afternoon. The seasonal peak power demand occurs in July and August.

2. System Descriptions

This system is mainly designed for a self-supported operating mode, with the utility grid as power back-up in the case of a lack of solar energy. Apart from this operating mode the system offers the possibility of supply surplus solar energy into the grid or test different kinds of mains supply. In order to reliably fulfill a part of the required energy supply it is necessary to integrate an energy storage. The considered lead acid battery of 8000 Ah at 300 VDE especially designed for solar generator systems permits

- to bridge cloudy periods
- to supply energy during dark periods and
- to store momentaneous surplus energy for periods with lower insolation.

In accordance with the designed different operation modes the battery will be divided into two separate groups, switchable between two DC busbars.

The main stage of power conditioning is the inverter system. It is provided that a defined number of inverters can be switched in or out in accordance with the momentary load requirement. This circuit design is important so that each inverter can be driven in its maximum efficiency range and thus minimize losses due to load mismatch especially by part loading.

The fact that nine different wind generators are already installed on the Island Pellworm offers an ideal opportunity for a later connection of both these alternative energy sources and the study of their supporting operation.

The total photovoltaic power plant will be controlled and monitored by a process computer. The system available on the site and originally intended for wind generator monitoring will be used preferably.

3. Solar Generator

The solar generator will be made up of flat, glass-encapsulated modules with solar cells which will have a size of 10 x 10 cm^2 and will be of multicrystalline silicon. With the previous mentioned size of multicrystalline silicon cells the smallest module unit will have an area of around 0,13 m^2. The power output of each unit is 9,2 Wp at AM1 spectrum and at a cell temperature of 25° C. To build up a 300 kW solar generator 32760 solar modules are necessary. The estimated total area of the solar generator including spacing between the rows of mounting frames necessary to prevent shadowing is ca. 125 m x 135 m. The final solar power plant will include the consideration of following criteria:

- clear, logical construction (easy to handle)
- optimization between cabling losses and shadowing of the active cell area
- optimum utilization of the topographical conditions of the site and the integration of the solar generator in the landscape
- easy access for installation, maintenance and cleaning
- protective measures (e. g. against electric shock hazard, lightning damage, trespassers, etc.
- safety of operation
- low-cost structure

According to this criterias the solar generator will be of modular construction and the modules will be interconnected into strings. Similar like the battery the single groups of the solar generator strings will be switchable between the two DC busbars.

The major part of the solar generator will be installed on a structure with an optimized tilt angle. A certain amount of the solar generator modules shall be mounted on a two-axis-tracking device of the MAN system HELIOMAN in order to investigate the features of tracking systems as well.

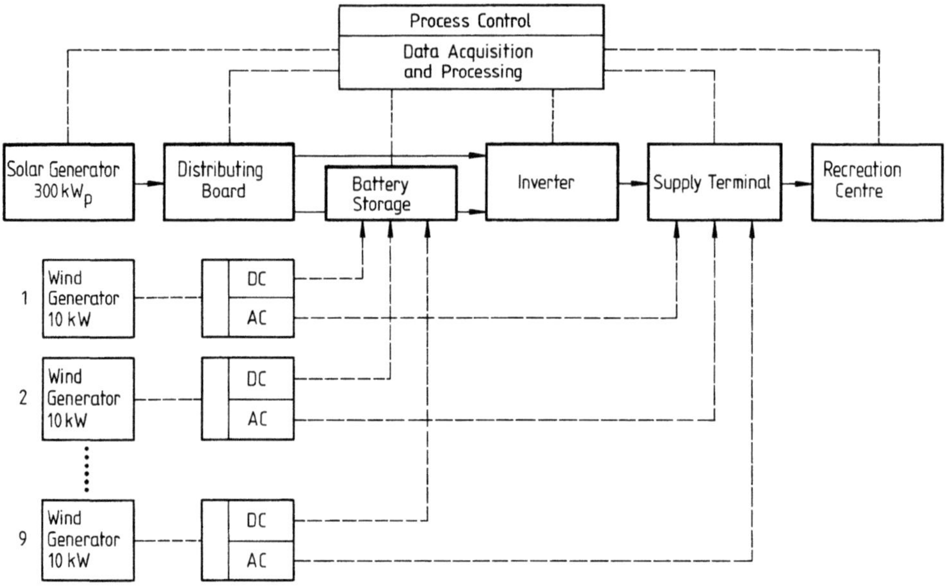

Fig.: 1 Block Diagram, Total System Approach

SOPHIA ANTIPOLIS 5 kW PHOTOVOLTAIC POWER GENERATOR FOR HABITAT

Daniel J. DENIS

OMERA - SEGID - Photovoltaïc Engineering

1. INTRODUCTION

This paper discribes the 5 kW photovoltaïc cell system that features several modes of solar energy collection, electricity storage in a 150 kWh accumulator and a specially tailored DC.AC converter. This power plant has been achieved to supply electricity to the tenants of a house just as the utility grid would do and to allow experimentation and selection of best suitable concepts and materials on a full size installation. Diagram is given in Fig. 1.

The whole power generator and dwelling house can be seen and visited, very close of the CERAM building, to the north of SOPHIA ANTIPOLIS, on the territory of BIOT.

This work is done for the C.E.C. (contract 468-78-1-ESF) and for COMES - (contract 79-01). It is kindly supported by CNRS/PIRDES (- R ans D solar energy program of the National Scientific Research Organization), - and by CNES - (French Space Agency.)

2. GENERATOR DESCRIPTION

The generator is installed on a hill culminating at 156 m above sea-level. Sun tracking arrays are located on the top while the house with the generator technical room near of it, lies a few feet lower. Large arrays - fix and seasonaly adjusted arrays - are installed near by.

2.1. Photovoltaïc arrays : Four types of PV collectors are used :

a) Two axis heliostat tracking sun both in azimuth and elevation simultaneously to maximize direct sun radiation. The installed peak power is 528 W.

b) Vertical axis heliostat following the sun in azimuth with a fixed tilt angle. The installed peak power is 528 W.

c) Seasonaly oriented arrays facing south. The tilt angle of this array is adjusted a few times a year. Peak power is 2112 W.

d) South oriented fixed arrays for which no adjustment is provided. Installed peak power is 2112 W.

The amount of energy produced by the four different types of collectors is recorded separatively to allow comparisons and to determine which of them is the most cost effective, considering installation costs versus amount of electricity produced.

2.2. Electricity storage

The system includes two main lead acid batteries which alternatively feed a 6 kVA inverter - one battery is under charge while the other is being discharged - Each main battery is able to store 75 kVA and is organized in 8 blocks of 48 V each.

In the discharge mode, the 8 blocks of the battery are connected in series (Total voltage between 355 V and 400 V). In the charge mode, each block is independantly charged by a dedicated section of solar array through a charge regulator.

2.3. Charge regulator

Charge regulators have been developed. Provision has been made for a dead zone in the regulation loop to avoid oscillations caused by a battery voltage raise if the load is disconnected. This hysteresis voltage varies depending on battery make and regulators have to be tuned accordingly before use.

2.4. 6 kVA inverter

A power 6 kVA inverter has been developed by SNIAS Toulouse - France for conversion of DC power from the battery into AC.

Its main characteristics are :

- Input : 360 V to 400 V DC - output : 220 V $\pm$ 2 % at 50 Hz $\pm$ 1 %
- distortion : less than 10 % for $0,8 < \cos\varphi < 1$
- nominal power : 6 kVA - peak power : 9 kVA
- efficiency : 94 % at P_n - 92 % at $P_n/10$
- Full protection against short circuits

2.5. Metrology unit

Metrology unit purposes are threefold :

a) It provides housekeeping informations such as charge/discharge currents, battery voltages.

b) It provides accurate data concerning the integrated incident solar flux for the 4 types of arrays, the energy produced by each of them, the energy charged in each battery, the energy drawn by the inverter and AC energy produced by the inverter.

These data will be processed to get the factor of merit of each array, the charge/discharge efficiency of the battery and fluctuation of power consumption for that typical house.

c) It triggers safety devices protecting the station e.g. heliostats arrays are automatically driven to the horizontal position when the wind speed is in excess of 75 kmh, battery discharge is interrupted and the plant goes into backup mode when battery voltage becomes lower than I.85 V per cell.

3. SPECIFIC PROBLEMS ASSOCIATED WITH USE OF SOLAR ELECTRICITY FOR HABITAT

Main problems are caused by fluctuation of load power drain. A tentative model for a house electricity consumption has been proposed by L.E.P. and a diagram of cumulated daily consumed energy has been reproduced in Fig. 2

Full load condition - 3 kW - lasts only 1 hour while most-of-the-time - 16 hours - power drain is lower than 200 W. Nevertheless a 6 kVA inverter is necessary to cope with motor starting up and with possible use of several high consumption of equipment at the same time e.g. use of a vacuum cleaner (1 kW) together with heating of water in the washing machine and other appliances.

The only constraint placed on the housewife is that she should refrain from using simultaneously the washing machine and the dish washer.

In any case, the inverter has been designed to be able to supply 9 kVA without any damage before current limiting circuits become active and reduce output voltage.

4. CONCLUSION

Use of photovoltaïc cells for habitat is a worst case application. This is due to the fact that the power consumption histogram is unfavorable that most power is consumed outside electricity production period and that long time autonomy is required. This pilot installation is therefore appropriate to bring to light most of the problems to be solved. Experience gained from this project is thought to be very valuable for other applications in which so many adverse conditions are unlikely to be found simultaneously.

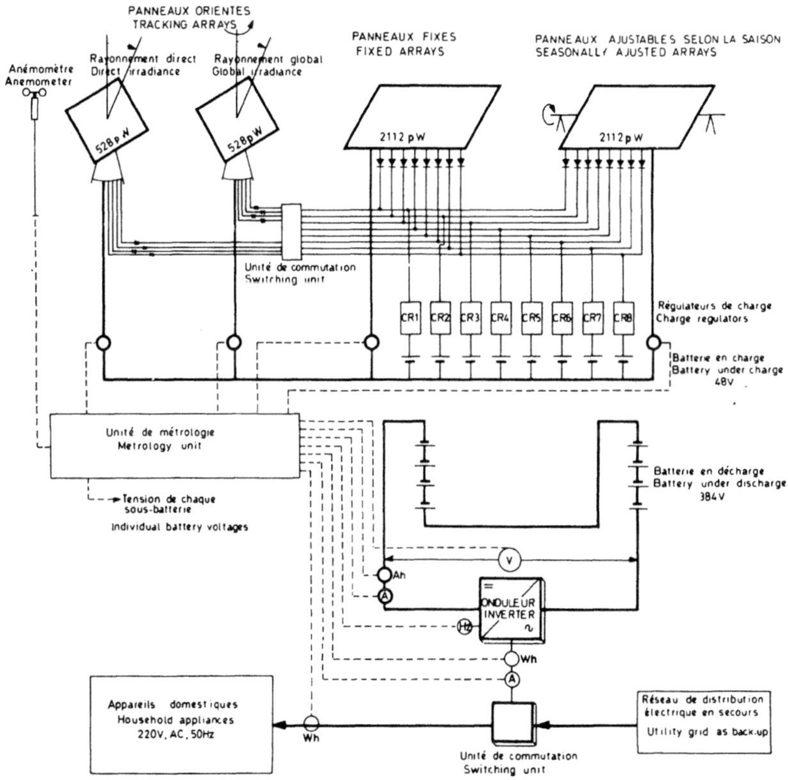

Les pointillés indiquent les connexions des appareils de mesure
Dotted lines show connections to sensors

A 50 kW PHOTOVOLTAIC POWER PLANT COUPLED TO THE 15000 V ELECTRICITY GRID

J. CHADJIVASSILIADIS
PUBLIC POWER CORPORATION - D.E.H.
ATHENS GREECE

J.L. CHENON and D.J. DENIS
Société OMERA-SEGID
ARGENTEUIL - FRANCE

Summary

Electricity powering of inhabited and isolated regions always implies a critical problem and the usual solutions do not bring very satisfying answers. Generally there are :

- Special connection to the national network
- Installation of small autonomous power stations
- The most common is the implantation of medium-size power plants for a few villages near to each others

In any case, the evolution of the cost price of the distributed kWh is unforeseenable.

A Greek-French consortium proposes to study the solution brought by a P.V. power plant of 50 kW in actual size. The site will be choosen amongst the numerous little villages in Cyclades Islands - GREECE.

After describing photovoltaïc panels arrays, the author exposes that the choice of the site will greatly influence the project. Two different conceptions are possible according to the fact that the P.V. station will be matched or not to another electricity network.

- Influence on the storage : one can consider that the connection to a mains grid is equivalent to an unlimited storage device.

- Influence on the inverter : the overcost for synchronizing on the mains makes it possible to cheaply improve the performances by continuous fixation at optimum point.

1) INTRODUCTION

The Greek space includes a large number of islands which, for the most part, are small. The only energy sources for these islands, at a very high cost, consist of the imported and transported liquid fuels.

Some of the islands which are close to the Greek mainland have submarine cable connection which is used for electrical energy supply from interconnected power grid.

Every year, there is an effort made within the electrification program to supply electricity to some of these islands with conventional stations and diesel engines.

The main advantage of these islands was known from the ancient times and from Greek mythology : it is the great duration of solar irradiance all over the year.

Because of these favorable conditions, modern technology could provide solutions to this energy coverage need in these islands as well as other remote sites of the continental part of Greece. The application of an experimental installation using solar radiation and photovoltaïc conversion could provide us with many useful conclusions for broader applications in the future. For such an installation, we can choose a representative island with its own electric energy system where the experimental photovoltaïc conversion unit would operate in parallel with the grid ; or we can install an equivalent station being the only way to power a group of houses.

In any case, it must be easily accessible and not too distant from Athens in order to be in better conditions to follow up with the installation work and the operation.

2) 50 kW PHOTOVOLTAIC POWER PLANT

A peak power of 50 kW has been choosen to provide a significant energy to the users. For instance it corresponds to the minimum needed by the whole island of KYTHNOS a winter day. Flat module BP X 47 C from R.T.C. - France has been selected.

The peak power of this module is 33 W and a total number of 1 500 modules is required. The modules are organized in sub-arrays with a bus voltage bar of about 160 volts.

The tilt angle of each sub-array is seasonaly adjusted. Three possible angles are envisaged corresponding to winter and summer for extreme positions, and to spring and autumn for medium position. Such a simple seasonal adjustment can bring as much as 10 % energy gain over fixed pannels.

3) ENERGY STORAGE

3.I) In case the selected site has not at all electricity power available, we shall set up regular set of lead batteries having a capacity able to face minimum of three days with completely dark sky. Considering the power of the station this battery will have big dimensions and will need special housing and maintenance.

3.2) So we should prefer a site connected to a grid as main power source. The maximum available 50 kW from photovoltaïc origin is always under the required level for a site as KYTHNOS e.g. - So, the mains may be considered as a storage for the power delivered by P.V. arrays because of economy in non-burnt fuel which remains available for the non-sunny period. When the regular power plants already exist on the site, such interconnected photovoltaïc system presents the following advantages :

- No cost for buying and housing batteries
- No maintenance for battery elements
- Increased lifetime for the whole system
- Lossless storage between input and output

- Utilization of the totality of the available energy (that is to say the grid is seen as an infinite capacity of storage)

4) INVERTER

4.1) Considering a new time an isolated site, without any other available electricity, the inverter will deliver a directly distributed voltage on low voltage line : 220/380 V triphased 50 Hz as usual in Greece.

Target in efficiency will be : 94 % at nominal power (p_n)

92 % at tenth of P_n

The no-charge losses will be less than 1 % of the nominal output power. This inverter should be an improved version of this one which is tested in the photovoltaïc house of SOPHIA ANTIPOLIS.

4.2) In case the site should already be equipped with local supply circuit, the main characteristics of the inverter will be the frequency. It has to be synchronous to the present mains.

To be always in the best efficiency conditions in the photovoltaïc conversion, the proposed device works on a closed loop. It does not use standard cell to measure sunshine. In order to avoid estimating working conditions throw false environmental conditions, it includes a microprocessor and a convenient algorythm which permanently check the intensity of the current and maintain the working point of the load at the maximum converted solar power.

Connection to the public grid will be done in 15 000 V-triphased - 50 Hz.

5) CONCLUSION

After implementation of such a power plant, il will be possible to know in actual conditions the real cost of one kWh. Actually, we can predict that in such a region where ground is very cheap, oil transportation very expensive and sunshine freely distributed, cost of photovoltaïc power stations will very soon be competitive.

Let us hope that it will stand to reason just when the first experimental results are available.

PHOTOVOLTAIC PROJECT FOR GIGLIO ISLAND

E. GAUTIER DI CONFIENGO - S.E.S. SpA

Manager Energy Systems

Technical data of the Pilot plant

The pilot plant described in this proposal has a 60 kW electrical peak power and the produced energy will be utilized in a reverse osmosis desalination plant and connected into the grid thus avoiding or minimizing any device of electrical storage.

These strategical choices are justified by the characteristics of the island which receives fresh water from the continent and has brackish water.

This utilization does not require any further storage devices due to the storage capacity of a water reservoir.

Our system can therefore be described by the following sub-systems:

- a set of silicon photovoltaic cells placed and interconnected in flat plate order, particularly studied in relation to the size of the plant (60 kW) and to the installation site in the sea weather, which is particularly demanding;

- a plate supporting frame which has to assure not only the installation and protection of the plates, but also economcy costs, and architectural compliance with the touristic landscape requirements of the site. Attention will be given so that the structure of the frame will be modular, applicable and suitable to the site, will require minimal energy input for production, will be transported easily and fitted to systems in different sites and with different features;

- an electrical subsystem to collect the produced energy and to convert it in the most suitable ways accordingly to the subsequent use (220 volts a.c. monophase, or a.c. 380 volts a.c. triphase).
  An optimization study will be made with respect to the electrical output of the system for different levels of insolation;

- an electrical interface with reverse osmosis and pumping system.

The system will be supplied with the necessary equipment for the recording of data during the functioning of the pilot plant.

Automatic functioning, easy maintenance, safety and reliability will be priority targets.

The great deal of experimental data will be recorded with the assistance and cooperation of the National Electricity Board (ENEL) and will represent a useful completion to similar experiences which are now being performed in other European countries.

The proposed project will comprise a theoretical study, so as to determine the optimizing conditions for the functioning of the system.

This study will mainly consist in the set-up of a mathematic model which will at first avail itself of meteorological and insolation data supplied by the existing observation stations.

As the work will progress and, above all, with the operative start-up of the photovoltaic system, it will be gradually possible to enter the calculus with more accurate data and the relevant feed-back will allow to match the previously planned data of the project in terms of confirming or possibly adjusting them, both on the technical as well as on the economical side.

On a first hand evaluation one could say the Isle de Giglio photovoltaic system is bound to work under fairly similar conditions as a project developed for Grosseto (Roselle) though this latter plant is differing in its project and utilization data.

By a rough comparison with Roselle one can expect that the yearly energy output of the system under average meteorological conditions will come up to 100.000 kWh.

Taking into account the foreseen price reduction for photovoltaic components and on the other side the equivalent increase for Diesel-oil price (Diesel-oil is the only possible gas-source at "Giglio"), such systems appear to be more and more competitive.

We wish to draw attention to the fact that the plan includes the connection with the electrical grid fed by the Diesel generator and also with an eolic system, planned, financed and installed by the Electricity Board. This latter connection will therefore not be on E.E.C. charges. Installing such an eolic system in a site with high winds like the Isle de Giglio will further add significance to the performance of the photovoltaic system.

Before drawing conclusions we think of some interest explaining more in detail the programme and the tasks of the Solar Consortium Giglio now on the way to be formed.

The Consortium includes public organisations such as the Region of Toscana and the town hall of the Isle de Giglio, who take on their charge together with the financial share not covered by the E.E.C. the liaison as well the optimization work within the whole project. For this purpose, the administrative and technical staff of the regional authorities will be particularly involved in the project.

The firms belonging to the Italian-German Consortium are:
Agip, Ansaldo, Dornier System, Galileo, Nuovo Pignone, OTO Melara, S.E.S. and Solaris.

CONVERSION AND STORAGE OF SOLAR ENERGY BY MEANS OF HYDROGEN

C. NOKERMAN, Director-General Manager E.N.I.
Association "GENSUN" (SGB-MHO-TE-ENI)

1. *GENSUN*

The association "Gensun" groups four leading companies of Belgium.

- *S.G.B., the biggest industrial holding company*
- *M.H.O., the biggest non-ferrous metallurgical refinery*
- *Traction et Electricité, the biggest engineering office*
- *E.N.I., one of the biggest electrical engineer and electro-mechanical contractor, who is the leader of the project.*

2. *Location and demonstration targets*

The project we are introducing by the EEC is a 30kW project, to be installed in a metallurgic plant, non-ferrous materials of MHO at Olen, near Antwerp, in Belgium, intended to demonstrate:

- *the use of low-cost original silicon solar cells on industrial site and in agressive atmosphere*
- *the use of very original electrolyse cells, to convert the current of the solar cells into hydrogen and oxygen, with very high efficiencies*
- *the ability to store easily the gaseous fuel hydrogen, without degradation of the stored energy on long term periods, without losses and without maintenance.*
- *the use of hydrogen for metallurgical operations namely the production of silicon*
- *the use of hydrogen for emergency and public lighting, for cooking, for heating and cooling, air conditionners and freezers.*

As shown on the mock-up, the solar cells will be installed on the roof of two industriel buildings, one of which is housing the existing elctrolysis installation.

3. *Topics of the hydrogen production*

3.1. *Silicon cells*

The silicon cells, 4" diameter, are produced by the process of the Belgian Flemish Catholic University of Leuven (KUL), ESAT Laboratory, under leading of Prof. R. Van Overstraeten.

It allows the use of solar-grade silicon of low resistance: 0,5ohmcm and characterise for a serigraphic process for doping and also for printing of the electrodes.

The efficiency of the modules is 7,5%

3.2. *Electrical feeding of the electrolyse cell*

To be made through a DC/DC convertor and/or a variable serie-parallel coupling of the elements of the electrolyse cell, working automatically to adapt the voltage to the characteristics of the cell.

3.3. *Production of gaseous hydrogen and oxygen*

The project will use the technics and special materials (menbranes and electrodes) developped by the

CEN-SCK (Center for Nuclear Energy) a State Research Institution of Belgium.

The research under development by the CEN has for target the use of very high current density, to minimise the cost.

Special membranes, using polyantinemic acid on a substrate, and special catalytic electrodes have been developped and patented, aswell as special electrodes still on small scale (28cm^2)

Using the same principles, the project will make the up-scaling of that system to bring the electrode surface to 314 cm^2 (20cm diameter).

Simultaneously the project will aim to get the highest possible efficiency, associated with a low production cost.

As result, the project incorporates an electrolyse cell operation under atmospheric pressure, at low temperature (40-60°) with a low electrode density current (0,2.. 0,3 A/cm^2), working completely static. The cell will be made for methyl metarrplate.

The target is to produce 1Nm^3 hydrogen for a consumption of 3,5kWh, with a effiency higher than 80%

4. *Storage*

The storage will be made at a pressure of around 1,05 bar, just high enough to feed the consumers.

The maximum hourly production of hydrogen will be 30/3,5 = 8,57Nm^3/hr.

With a gazometer of 2 0Nm^3 the storage capacity will be 2,33 hours full production or 59kWh thermic.

Such a gazometer is still easy to handle and easy to install.

The low pressure choosen in this case is intended to obtain the highest possible global efficiency.

5. *Utilizations*

The use of the stored hydrogen is very wide. It can be used for lighting, cooking, heating,cooling-air conditionning and freezing.

Hydrogen is the ideal gaseous fuel, burning without any pollution and with a very low noise level.

We intend to demonstrate that :

a) long period hydrogen storage is possible without any losses or alteration, and practically without maintenance.

b) hydrogen is the only fuel permitting to produce easily very high levels of temperature, to feed, for example, torches for metal cutting.

c) hydrogen is a fuel permitting to develop suddenly, without limitation, very high peaks of energy, aswell as a metered use.

d) the regulation of the consumers will be very simple, and well adapted to the users in foreign countries abroad.

RENEWAL OF RURAL LIFE BY PHOTOVOLTAICS

A PHOTOVOLTAIC PILOT PLANT FOR RURAL COMMUNITIES

JACQUES RECHOU "Research-Development" and DOMINIQUE MERCIER "Solar Systems"
Société des Moteurs LEROY-SOMER - 16015 ANGOULEME (FRANCE)

GENERAL : The present project is mostly aimed at rural communities development.
As a matter of fact, all services usefull for this type of rural living conditions are brought together within the same installation :

- Electric power supply to houses
- Street lighting
- Water pumping
- Milkery with milk refrigeration
- Mechanics repair shop
- Optionally small solar powered telephone commuting system

This installation, located in CORSICA will supply a hamlet of 7 families. It can supply, as it is, secondary centers in Developping Countries.

DESCRIPTION

• POWER PLANT

- The plant has been designed for 52 kw peak power. However, if needed, 44 kw only could be installed during a first step and would be almost sufficient for supplying the present inhabitants. But the optimum size of 52 kw will have to be installed shortly afterwards.

- FRANCE PHOTON silicon cell fixed panel arrays will be installed (no tracking - no concentration) in a field oriented fully southward within a few hundred meters from all users.

- A set of lead acid, specially designed electrical batteries made by OLDHAM will be able to store 250 kWh which will provide 3.5 days under winter conditions at 80% discharge.

- Static dc-ac 50 kVA converters, made by the french company SNIAS will convert the 192 volts DC into 50hz 220/380volts AC. Efficiency has been guaranteed between 90% at full load and 85% at 25% load.

- The total energy consumption has been computed at :

  • 75 kWh per day in winter :
    i.e. 45 kWh for home uses (7 farmers and their flocks)
    10 kWh for water pumping
    5 kWh for street lighting
    15 kWh for milkery and repair shop

  • and 160 kWh per day during summer :
    i.e. 100 kWh for home uses (15 families) 7 farmers and 8 summer visitors
    20 kWh for water pumping
    5 kWh for street lighting
    35 kWh for milkery, refrigeration of the milk and repair shop

  Additional panels will provide 300 watts 48volts DC and 66 watts 12 volts DC power for the telephone commuting system

OVERALL COST : The 52 kW installation has been evaluated at an overall amount of 1.6 Million of U.S. Dollars.

Including :

- Photovoltaïc array	850.000 US $
- Structures and interconnections ..	130.000 US $
- DC/AC inverter	100.000 US $
- Storage batteries and regulator ..	360.000 US $
- General design study	55.000 US $
- Instrumentation and miscellaneous.	105.000 US $

ECONOMICAL INTEREST OF THE PROJECT

In itself, the present project represents a vital need for this area of CORSICA.

As a matter of fact 800 people used to lived in RONDULINU in the years before 1880.

Progressively, due to the lack of basic facilities, all families migrated, and only 7 families remain breeding domestic animals.

It is the task to local autorities to bring to these families electric power so to incite them to stay and, if possible to bring other families back.

Help from local corsican autorities is established, as well as the acceptance of the corsican population.

DEMONSTRATION VALUE OF THIS PROJECT

The site is overlooking the SAGOGNE gulf.

Located within one hour drive from AJACCIO airport, fifteen minutes from CARGESE (where a summer school is active) through a hard top road, the site is easy to visit.

Furthermore, the town of CARGESE is very dynamic and one can think that all facilities to visitors will be brought by local autorities.

All utilisations of photovoltaïc electricity will be concentrated within a few hundred meters and will forcefully demonstrate the usefullness of this source of energy and the readiness of its applications to rural populations.

One must realize that in remote rural areas of developping countries where electricity had never been supplied, the power consumption per family is much lower than the average rural european consumption.

It can be estimated that in such countries, average needs are roughly 1 kWh per day and per family.

Consequently the present project could be applied, as it is, to secondary remote centers of around 150 families (i.e. 1000 to 2000 inhabitants).

A 60 kW PHOTOVOLTAIC RURAL ELECTRIFICATION FOR THE ISLAND OF LA REUNION

SERI RENAULT INGENIERIE	(F)
JEUMONT - SCHNEIDER	(F)
LA RADIOTECHNIQUE-COMPELEC	(F)
VARTA A.G.	(FRG)

1. INTRODUCTION

The above mentioned consortium is proposing the electrification of the Cirque de Mafate in the Island of La Réunion (French Overseas Department) by a 60 kW photovoltaïc system. About 400 people live in this 15 km long 6 km wide rift valley surrounded by a barrier of mountains. Too isolated to be connected to the island utility grid, this rural population, subsisting on cattle and food crops, has been slowly decreasing since 1850. The improvement of the living conditions is essential to restrain the rural exodus to the coastal towns. The objective of this project is to demonstrate photovoltaïcs as a suitable and reliable energy resource for such a remote and rural population.

2. SITE DESCRIPTION

The closed little world of Mafate contains just 130 homes, 4 schools, 5 tourists shelters and a well maintained footpath network at an average altitude of 1,200 m. The sunshine duration is c.a. 2,500 hours evenly distributed throughout the year (170 hours in December; 220 hours in July). The average temperature varies within the range of 15 to 20°C.

These favorable weather conditions are important for the capability of this stand alone PV generator to cope with the real demand profile.

3. SYSTEM PROPOSED

The proposed 60 kW PV system will be installed by SERI RENAULT Ingénierie. The installation will make use of helicopters as did the 5 kW PV generator project at the refuge des Evettes.

Jeumont-Schneider which has a technical staff office at Saint-Denis (La Réunion) will be in a position to provide some assistance. The maintenance will consist of routine checking visits.

The homes in the Cirque de Mafate are mainly collected in two places "Aurère" and "La Nouvelle" which are about 5 or 6 km away from each other.

In order to minimize the distribution costs and taking advantage of the modularity of PV generators we decided to split the system into two subgenerators of 30 kW. From each subgenerator a wooden truss line will be erected by the Direction Départementale de l'Agriculture to distribute 220 V ac electricity to the homes and to the communal facilities of "Aurère" or "La Nouvelle".

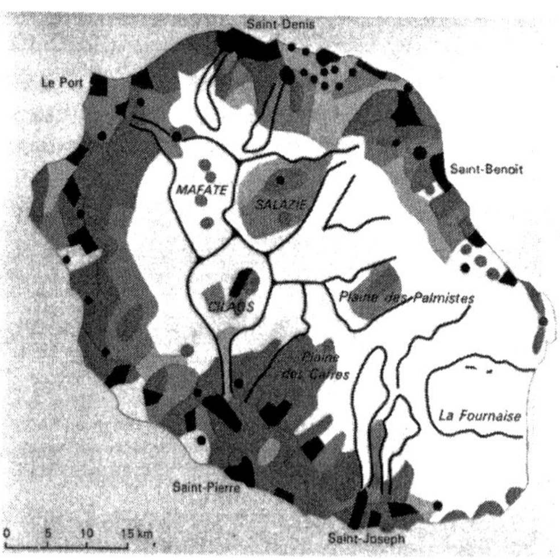

Island of La Réunion : population density showing the "Cirque de Mafate"

4. DESCRIPTION OF THE SUBSYSTEMS

Each of the 30 kW subgenerators features :

- 910 modified modules (BPX 47 C) from RTC. They will be beld by simplified supporting structures that allow fast mounting. Wiring is specially designed to minimize work on site by using the truck industry's fast cabling techniques.
- A device develoved by SERI RENAULT Ingénierie modifying the Series/ Parallel configuration of part of the modules, will track the maximum power voltage and increase the energy produced yearly by about 15 %.
- The storage - using OPzS stationary batteries by VARTA consists of 150 elements of 1,500 Ah each. The storage efficiency will benefit from the large capacity installed which will be sufficient for the system to be autonomous for 3 days.
- An output inverter, specially developed by JEUMONT-SCHNEIDER using new transistors and the Pulse Width Modulation Technology, and fed by a 330 V dc voltage, supplies 220 V single phase ac with a global efficiency of over 90 % for a load ranging from 10 to 100 %.
- A control command panel.
- The building and the utilities (lightning protection, fence,...).

Each subsystem is selected according to its reliability and automatic performance which is necessary for operation in an isolated area.

5. PERFORMANCE ASSESSMENT

The subsystems (batteries with low self discharge, maximum power tracking device, inverter) as well as the energy management (low current intensities) and the safeties selected by SERI RENAULT Ingénierie make it possible to obtain an excellent global energetic efficiency for the system.

Taking into consideration the estimated daily irradiation and the global efficiency, it is anticipated that the system will supply about 150 kWh a day (avg).

THE 5 kW GENERATOR AT THE ALPINE REFUGE "LES EVETTES"

L. SELLES - B. AUBERT
Société d'Etudes et de Réalisations Industrielles (SERI)
RENAULT INGENIERIE - Bois d'Arcy (F)

1. INTRODUCTION

This 5 kW photovoltaïc generator project was undertaken within the framework of the first four year (1976-1979) Solar Energy R and D programme of the Commission of the European Communities to demonstrate that existing photovoltaïc system technology was capable of providing electrical power for remote houses.

By the End of 1978, SERI RENAULT Ingénierie after consulting the E.C. Commission, the "Région Rhône-Alpes" and the "Club Alpin Français" has chosen the "Refuge des Evettes" as the most suitable site for this project. The 66 bed hotel is not tied in to the "Electricité de France" national distribution network. This circumstance, the real operating conditions, in addition to the difficulty of access (supply by helicopters only) were particularly suitable for the demonstration.

The installation of the system was made during the Summer of 79. Due to late snows this year, the system was put into operation on August 1., 80 and ran properly till the end of the season (September 22.).

2. THE 5 kW STAND ALONE GENERATOR :

SITE	Refuge "Les Evettes" - Commune de Bonneval S/Arc - Savoie - France Latitude : 45° 22'N - Longitude 4° 46'E, Altitude 2,600 m. Open from Easter to end to Summer.
PHOTOVOLTAIC POWER	ELECTRICAL CHARACTERISTICS : Peak PV Power : 5 kW; Voltage : 48 - 70 V dc Daily generated energy : 25 kWh (avg) (32 kWh peak; 8 kWh min) COMPOSITION : RTC Model BPX 47 A : 452 modules of 11 W each supported by 38 galvanized welded iron structures. Tilt : 20°/horizontal; overall height = 50 cm; total area 70 m2.
STORAGE	BATTERY : 120 lead-acid elements of 600 Ah each, 48-56 V, 3,000 Ah in 10 hours (144 kWh). AUTONOMY : 5 days without direct irradiance at usual demand-load.
POWER CONDITIONER	CONTROL AND REGULATION UNIT : a) Storage capacity management by adjustable thresholds; b) Control command of system - 3 band time schedule programmer; c) Safeties. INVERTER : 3 kVA, equipped with thyristors 50 Vdc - 220 Vac (50 Hz). Efficiency : 80 % at full load, 67 % at mid load.
OTHERS FEATURES	LIGHTNING PROTECTION : Radioactive head rod and earthing of all metallic structures. DATA RECORDING : 12 channels, 24 Vdc, paper chart recorder running permanently for : irradiance, temperature, current, voltage, power,...

3. DESIGN OF THE SYSTEM

The key point of this work was the global optimalization of the system, taking into account both solar irradiation and the users'needs.

The system is designed around the battery which is used to decouple irradiance and utilization. The battery was set up to operate in the conditions in which it could work most efficiently. Likewise, in order to increase the overall energetic efficiency of the inverter a time schedule was imposed to limit its operation at no load.

The self reliance of the operation and the minimal maintenance required appeared to be very important criteria for such a stand alone installation which has to be operated and maintained by the Refuge keepers who have no academic technical background.

4. EXPERIMENTAL DATA

August 80. Solar Data : Maximum (at noon) global irradiance in the array plane varies from about 200 W/m2 up to 1,100 or even 1,200 W/m2 when the weather was variable with scattered clouds.

Measured daily global irradiation was 7 kWh/m2 (avg). Outside temperature was in the range + 5°C, + 20°C during the day (-5°C, + 5°C at night).

August 80. Performance Data : On the brightest days the output dc energy of the array, varies from 28.2 kWh up to 33.6 kWh. The losses associated with the dispersion of characteristics and with the disadaptation effect are evaluated between 33.4 % and 20.9 %.

Measurement of inverter efficiency meets the constructor's specifications, its cumulated energetic efficiency was 65 % on the average.

The demand profile average was 10.5 kWh (max 17 kWh - min 4 kWh).

5. CONCLUDING REMARKS :

Overall efficiency of future larger PV generators could be increased by 10 to 20 % by means of a voltage matching device and then increased by 20 to 30 % by use of an inverter fed by a higher battery voltage and which is more efficient at widely varying loads. These two components that were lying in a R & D stage when the generator was designed, will be in a few years commonly available to system designers.

More than 80 % of the energy used passed through the storage battery. The importance of a large storage capactiy and of its state of charge management must be emphasized.

Although the improvement of the living conditions was deeply appreciated, the users of the system (the Refuge keepers) found it difficult to accept the constraints of the 3 bandtime schedule on the use of inverter. In view of the demand profile encountered, the time schedule was finally relinquished without altering design adequacy.

The extreme enthusiasm of hikers and alpinists was very encouraging and augurs well for future systems.

POWER SUPPLY FOR A SOLAR WATER HEATING INSTALLATION

J. P. GOMIS J. C. VIALART

Executive Vice President System Department Manager
of PHOTOWATT INTERNATIONAL SA.

ABSTRACT

The application consists in using a small solar photovoltaic station installed in a holiday resort, to supply the electric part of an existing solar water heating installation (pumping, control unit), and be used as an emergency power source for the main safety needs of the village in case of general failure of the electric distribution.
In the field of exploitation of solar energy, this application carries out the synthesis of both thermal and photovoltaic ways. To be used in the leisure industry, the project comply with the usual requirement of autonomous power supply existing in these activities. In other respects, this application allows to test a type of simple and handy equipment which can be considered as a possible "village power source" for developping countries.

APPLICATION

. Objective of the application

The thermal solar energy for production of hot water (sanitary use) is already operating and will probably be more and more used, due to its profitability.The thermal installations for water heating need a part of the energy under the electric form; it is a small part of the total power needs but essential for powering the pumps for water circulation and control units.
The objective of the proposed application is therefore, to produce by the way of the photovoltaic conversion, the electric energy necessary to a COMPLETELY AUTONOMOUS OPERATION OF A SOLAR WATER HEATING INSTALLATION FOR HOLIDAYS RESORTS VACATION CAMPS.
In other respects, a storage of electric energy being necessary to insure a sufficient no sun autonomy, it is interesting to be able to use this energy as an emergency source, in case of total failure of the general distribution of the village. This source can supply indispensable needs essentially destined to safety requirements.

. General description of the application

The photovoltaic power source for holidays village is therefore used for the two main following applications :

. supply electrical power to pumping and control units of the solar water heating installation,
. emergency power source for the village.

The various functions for the water heating installation are summarized in the following way :

- water circulation in the primary circuit,
- water circulation in the secondary loop,
- water treatment unit (against hard water),
- monitoring safety and signalisation unit.

. Proposed user and site

Company .... Club Méditerranée
Proposed site .... Village "les Boucaniers", Pointe à Pitre, La Martinique, french overseas department.

. Particular characteristics of the site

WATER HEATING INSTALLATION : 950 m2 of thermal panels, about 80 m3 of hot water per day.

Primary circuit : 3 pumps 1.5 kW each, 2 pumps operating together, one in reserve; use of the pump during the sunlight period, i.e. an average of 12 hours per day
daily consumption .... 36 kWh

Secondary loop : 3 pumps 0.55 kW each, 2 pumps operating together, one in reserve, use of the pumps during an average of 18 hours per day, daily consumption .... 20 kWh

Water treatment unit : injection of products of treatment according to the consumption of water, a unit of 1 kW operating during an average of 15 hours per day,
daily consumption .... 15 kWh

Control and safety unit : total power of 1 kW used as a permanent equipment, i.e. 24 hours per day,
daily consumption .... 24 kWh

The total daily energy consumption is therefore taken at a level of 95 kWh/day.

WATER ENERGY SOURCE FOR THE VILLAGE : the safety needs in case of emergency (total failure of the general distribution network of the village) are the following :

. minimum general lighting,
. pumping system of used water and rain streaming water as a part of the village is easily flooded in case of heavy rain.

DESCRIPTION OF THE PHOTOVOLTAIC SYSTEM

. General presentation

The proposed photovoltaic power station is composed of :

. a solar array of 30 peak kW power providing energy collection,
. a 300 kWh electrochemical storage with protective devices,
. DC/AC conversion with two static converters,
. simple interface with the user with manual commutation of the load either on solar station or grid; automatic switching of "emergency" services on the photovoltaic system of utility disapears,
. data logging : to display and account the typical datas of the system.

. Energy collection

Photovoltaic module manufactured by PHOTOWATT and LUCAS will be used for collecting and converting solar energy. The cells used in these modules will be exclusively made by PHOTOWATT.
The general tentative data for PHOTOWATT modules is :
. circular monocrystal or semicrystal 100 x100 mm cells,
. encapsulation laminated type (PVB or EVA),
. superstrate : tempered glass,
. substrate :aluminium/polyester composite,
. frame anodized aluminium,

. redondant cells to cell and string to terminals interconnections
. junction box on back substrate,
. nominal power 35 or 70 watts,
. minimum power 30.5 or 61 watts,
. voltage at P max. 16 ± 1 volt or 32 ± 2 volts.

The modules type L 1219 manufactured by LUCAS Company, using semicrystal cells from PHOTOWATT, have the following general characteristics :
. peak power : 35W,
. encapsulation : laminated type,
. superstrate : hardened glass,
. substrate : atmospheric conditions resistant,
. dimensions : 1070 x 40 x 98 cm.

. Solar array arrangement

The photovoltaic field is composed of 30 parallel strings of 28 modules, serial mounted, i.e. a total peak power of about 30 kW with a nominal voltage of 400 V.
The tilt angle is 15°.

. Power conditionning

This operation is based on simples and hardy components providing to the station a good reliability.
Two mains functions are performed :
D1. control of DC energy and storage,
D2. DC/AC conversion.

Principles of working :

D1. the control unit for DC energy provides functions equivalent to those of the charge controllers of standard photovoltaic systems.
The storage battery is connected to the solar array, the charge controller avoiding any overcharge of the battery as well as loosing of water. The control is serial type, providing a limitation of the current from the array when the storage battery has reached a sufficient level of charge. The storage is an electrochemical one. The total capacity is 300 kWh, by the way of 175 lead acid cells, low antimonious rate type with a self discharge lower than 5% per month.

D2. DC/AC conversion
Two possible versions have been considered for this operation :
. used one converter capable of delivering the peak current, but really under-charged during the largest part of the operating time,
. use several converters allowing each of them to fit better with the requirements.

The second solution has been chosen It improves the reliability of the system by a possible commutation of the load on either one converter or the other in case of failure of one of them. This solution allows also the function of the emergency source in case of failure of the grid.
A static converter of 7.5 kVA supplies the pumps of the primary circuit The second converter of 10 kVA insures the supply of the other functions and is able to deliver the necessary power for the emergency functions. The wave shape delivered by the converters is well adopted to the utilisation and the control acts in order to avoid no-charge losses of the converters.

. Emergency/safety functions

EMERGENCY : in case of failure of the mains, priority functions must continue in operating : pumping of used water, minimum lighting of the village. The photovoltaic station can insure this service with a priority, if necessary, on standard functions for water heating installation. The come back to normal working takes place at the end of the failure.
SAFETY : the status of charge of the battery is permanently checked for both standard and emergency functions. An alarm device informs when a first level of abusive depth of discharge is reached. At a second level a load shedding of the converters can be make to protect the battery.

Data logging

The simple function shows the instantaneous status of the station by a data logging at various levels and also allows a general measurement by accounting of the different characteristics.

. User interface

The interface is designed in order to keep the existing distribution structure. It allows to easily and manually come back to existing energy distribution to avoid any trouble for the user as well as to preserve both possibi lities of supplying.

. Efficiency

Taking into account the line losses, the efficiency of the storage battery (more than 85%),the average efficiency of the DC/AC converters (more than 85%), various parameters such as modification of the working point of the modules due to temperature charges, the daily average of the delivered energy must be considered as more than 120 kWh while the solar array itself deli vers an average of 170 kWh per day.
The actual standard consumption of 95 kWh per day allows a safety factor of about 20% signifying a quite complete autonomy of the station.

PHOTOVOLTAIC PILOT PLANT FOR THE WATER SUPPLY OF THE TREMITI ISLANDS

F. FONZI

Adriatica Componenti Elettronici S.p.A.

67039 Sulmona

Strada Statale 17

ITALY

1. Introduction

The purpose of the realization is to render the Tremiti Islands self-sufficient as far as drinking water is concerned; avoiding the actual expensive periodic transportation through water-supply-ship.

The design plant is able to produce the necessary drinking water from seawater with 80 kWp of the photovoltaic generator for the yearly overall needs for the local inhabitants or to cover the needs of the S. Nicola Island including the conspicuous summer touristic presence.

The overproduction of photovoltaic electric energy will be used especially in the winter time for the illumination of the islands.

2. Process choice

Among the various processes of desalination, the reverse osmosis process seems to be the most suitable to be associated to the photovoltaic system having less energy consumption per m^3 of drinking water (10 kWh/m^3).

The process consists of desalination of seawater through semi-permeable membrane (*), passed by the water at a pression of 60 bar.

The residual saline content drops to about 100 p.p.m.

For every m^3 of drinking water it is necessary to treat not less than 3 m^3 of seawater.

With the photovoltaic generator of 80 kWp it is possible to obtain during the summer months 30 m^3 of drinking water per day.

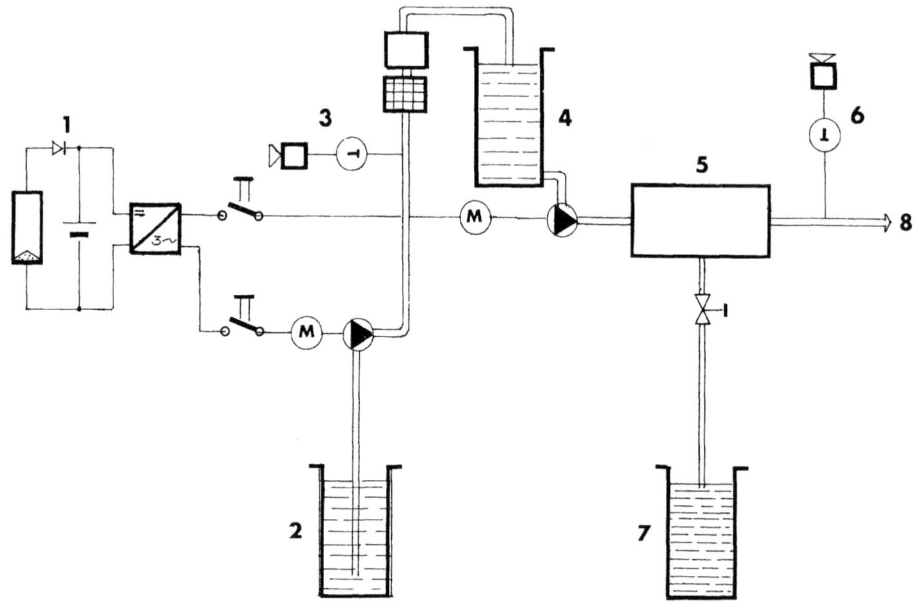

1. Solar generator with batteries and inverter
2. Seawater pumping
3. Seawater pretreatment
4. Storage of pretreated water
5. Reverse osmosis desalination unit
6. Post-treatment of drinking water
7. Brine return
8. Storage of drinking water

(*) see diagram

3. Design Data

Having in mind the local conditions of insolation of the Tremiti Islands (peak power 940 W/m^2) and the conversion efficiency of the photovoltaic cells (12 %), the collecting surface of the panels should be made equal to 710 m^2. The effective electric energy, deducting losses (batteries, inverter) is equal to 84,500 kWh per year, with a minimum and maximum monthly value of 4,100 kWh (December) and 10,500 kWh (July), respectively.

Such energy allows a yearly production of about 8,500 m^3 of drinking water, which is enough to cover the needs of the local inhabitants of all the Tremiti Islands (about 250 persons), when assuming a daily consumption per person of about 90 lt., or the overall needs (residents and tourists) of the S. Nicola Island, where the plant will be placed. Since the yearly overall consumption of water of the islands including the tourists is equal to about 60,000 m^3, it could be necessary, for the purpose of self-sufficiency, to install a second desalination plant fed by a 300 kW diesel-electric generator working from May until September to meet the peak summer water requirement. For the storage of the water produced by the plants the existing reservoirs are sufficient (capacity 17,700 m^3).

4. Operating conditions

We foresee the permanent operating condition of the basic photovoltaic plant for 8h/day, whereas the electric diesel plant would operate for 24h/day, only during the summer months. For the basic photovoltaic plant it is necessary to provide a limited storage of electric energy of lead-acid stationary type batteries to assure a sufficient continuity of the process. We foresee a 6-hour storage (equal to 1,000 Ah at the voltage of 220 V DC), assuming that there should be an absence of sun for two consecutive days, the running of the basic plant is assured by the diesel-electric group.

The basic plant foresees a conversion unit made up of two parallel converters of 30 kVA 220 V DC, for the powering of the various pumps (lifting, compression, reactive dosage).

The lifting seawater pump (5 kW) will be activated during the central hours of the day (winter time from 11.00 h to 14.00 h, summer time from 10.00 h to 16.00) so as to exploit the peak power of the photovoltaic generator.

5. Characteristics of the photovoltaic plant

- Silicon panels will be used, practically monocrystalline (peak power AM1 = 120 W/m^2)
- The electronic sub-system will allow for the following performances:
  - a) to regulate the charge of the battery according to the procedures suggested by the manufacturers
  - b) to obtain the maximum power transfer to the load, depending on the temperature and on the incident energy.
- The panels will have a fixed N-S position, while the inclination can be manually adjusted so as to collect the maximum radiation.

THE EPSILON PROJECT

European Photovoltaic System Installation in London

P.R. Wolfe, Lucas Energy Systems Ltd, England

For photovoltaics to be a major contributor to European electricity generation at any time in the future, it must be viable more widely than in remote areas. An essential part of assessing its future must be to study performance in urban areas. Europe does not have limitless unneeded space so designs will need to integrate solar arrays into existing buildings.

The Epsilon project has been planned to provide vital experience of urban solar systems.

The intention is to install a 30 kW array on the roof of a television station in central London. The array size has been calculated to provide sufficient power to meet the yearly average load of one studio in the complex. The system is also planned to include a significant storage capacity and could double as a atandby power supply for the entire building. It would provide backing electricity for the whole complex on full load for ¼ hour, or on emergency load for 4 to 20 hours per day, depending on the time of year.

This paper considers the specific objectives of this project and notes some of the ways of approaching these aims, and lessons to be learnt from them. The generalities will not be covered, nor will many of the topics common to most of the projects of this EC programme. The quest for properly engineered systems with good overall efficiency is taken as understood.

The design and development work required for this project will of course be undertaken during its implementation.

This paper and the draft proposal to which it refers should not be considered to offer final solutions on the approach, merely to suggest areas to be explored during the project.

1. URBAN INSTALLATION

1.1 Design for built-up areas

In general urban applications it will be necessary for solar arrays to be roof mounted. This will require designs to be architecturally suited to their environment. This particular project on a highly prestigious site on the Thames in central London will need to spend particular attention to this.

1.2 Design for roof mounting

Fixed flat plate solar arrays must generally face the equator for maximum output. On existing buildings this is not always possible or easy. The effects of shadowing from neighbouring sites and of the orientation of the roof must be considered. In this particular project the site poses its own individual constraints. Computer analysis will be needed to calculate the shadow profile on the roof area to find the optimum location. A programme has already been devised to perform this and early results are shown on the site plan in Figure 1 overleaf.

1.3 Roof-top installation

Since the area available for the best insolation on roof tops may be limited, it might be necessary to employ high density modules, or mobile arrays to avoid shadowing. The fixing of arrays onto roof tops will also need new development work different from ground mounting.

2. URBAN OPERATION

2.1 Environmental effects

The atmosphere conditions in urban environments will be very different from the remote locations where the vast majority of photovoltaics installations have hitherto been made. This will effect not only the pattern of direct and diffuse radiation, but also probably the deterioration with time. Only the continuing metering of a system of this size and importance will identify the extent of these effects. Monitoring will be considered an important part of the project.

2.2 Popular acceptance

It must be amongst the principal objectives of the whole programme to bring the potential of photovoltaics to public attention. This project, with its prominent site, and a user who regularly broadcasts to millions of viewers would give a solid boost to this objective, of particular value to a nation without strong official commitment to photovoltaics.

3. INDUSTRIAL LOAD MATCHING

3.1 Temperate Climate Performance

The extreme seasonal variations in sunshine in the UK make it impossible to find many loads with good correlation to solar availability. The use of a system to provide standby power to the complex in addition to meeting the specified load is therefore of considerable value.

				91	91	91	91	91	91	91	91	91	91	92	94	94	94	96	96	94	93	94	94
				90	90	90	90	90	90	90	90	91	91	91	91	91	91	91	91	91	94	96	94
				87	87	87	87	85	85	85	85	80	67	**	**	**	**	**	**	**	**	**	**
				85	85	85	83	83	80	78	75	67	51	**	**	**	**	**	**	**	**	**	**
				83	83	80	80	77	75	74	67	62	51	**	**	**	**	**	**	**	**	**	**
				80	78	77	75	75	72	67	62	62	50	**	**	**	**	**	**	**	**	**	**
				77	75	75	74	72	67	63	62	51	49	**	**	**	**	**	**	**	**	**	**
				75	75	74	69	67	64	62	62	51	48	**	**	**	**	**	**	**	**	**	**
				74	72	69	67	64	62	62	51	51	48	**	**	**	**	**	**	**	**	**	**
				75	72	70	66	62	62	62	51	51	48	**	**	**	**	**	**	**	**	**	**
				76	72	70	67	64	64	51	51	51	48	**	**	**	**	**	**	**	**	**	**
				75	73	70	69	69	69	58	56	54	48	**	**	**	**	**	**	**	**	**	**
				76	74	72	72	72	72	62	62	61	59	34	26	26	26	28	28	28	28	33	46
				75	74	75	75	75	65	65	65	63	61	49	37	31	30	30	33	35	38	46	62
				75	75	75	75	75	65	65	65	64	61	54	45	42	38	38	39	40	46	51	62
				75	76	76	77	69	67	67	67	65	64	58	54	49	47	45	45	48	52	51	63
				77	77	77	77	69	69	70	70	68	66	62	60	54	53	50	53	55	54	64	65
				79	79	79	74	72	70	72	73	69	68	64	62	58	57	58	58	59	56	66	68
				80	80	80	74	74	73	73	72	71	69	69	64	64	61	63	62	60	69	68	70
				80	82	82	77	75	73	75	74	71	71	70	67	67	66	67	65	62	70	69	72
				83	83	77	78	77	75	75	75	73	73	73	71	70	71	67	65	75	72	70	73
				85	86	82	82	80	79	79	77	77	75	75	73	74	73	70	67	75	75	72	74
87	85	86	86	88	82	82	82	81	79	80	78	77	77	77	74	75	73	70	67	77	75	76	76
88	86	88	88	88	82	83	83	81	80	80	78	79	79	78	76	75	73	72	80	77	76	77	76
90	90	90	90	86	85	85	86	84	85	85	83	83	83	83	80	77	77	75	80	77	77	77	81
90	90	90	91	87	85	86	87	85	85	84	85	85	84	83	80	79	77	83	82	80	77	79	81
90	90	91	91	87	86	87	87	85	85	86	85	85	85	83	82	81	80	86	85	83	79	81	83
91	91	91	88	88	87	87	87	87	89	88	88	89	87	85	84	83	88	88	85	83	81	85	87
91	91	92	88	89	87	87	87	89	89	88	89	89	87	85	84	83	88	88	85	85	84	85	87
91	92	89	89	89	87	87	89	89	90	89	89	89	87	87	87	85	90	90	86	87	86	87	89
94	95	91	91	91	89	89	89	90	92	90	89	89	87	87	87	91	91	90	88	87	88	88	89
95	91	91	91	91	89	89	90	90	92	90	89	89	87	87	87	91	91	90	90	87	89	90	91

LOCATION OF OFFICE TOWER BLOCK

Flat roof area 3 storeys high

Table 1 Computer analysis of shade on roof area from tower. % annual available sunshine

SELFSUPPORTING PHOTO-VOLTAIC SUPPLY SYSTEM FOR THE SCHOOL OF MARITIME STUDIES ON THE ISLAND TERSCHELLING, HOLLAND

By: R.P.M. Sonneville
Holecsol Systems B.V., Eindhoven
a company of Holec N.V., Utrecht

Summary

The system provides about 95% of the electrical energy demand (ca 32000 KWh) of the school and about 50% of the boarding school (ca 40000 KWh). Local weather-circumstances (sun-radiation + wind) are used to its optimum by applying a photovoltaic solar-generator in combination with a wind-generator. For 2 reasons the system is connected to the public grid, in the first place to guarantee a completely reliable energy-delivery to the user, secondly to increase the use-factor of the system.

1. INTRODUCTION

As you all know the CEC invitation to prepare proposals for the 30 - 300 KW pilot projects includes already some important technical aims. Especially mechanical, wiring aspects of the solar-generator and the sub-system power-conditioning has to be studied carefully in order to decrease costs.

The storage-subsystem and the combination with an other power-set are important items equally.

Concerning the choice of the location (user) we were guided by the following considerations (Sheet I Geographical Situation).

- After a test-period of one year the system should be transferred to the user.

- The user and/or local authorities will put a certain surface (ca 1500 m^2) at our disposal and support the installation of the system in every respect.

Besides these criteria Holecsol and AEG Telefunken will pay a lot of attention to some additional aspects.

2. ADDITIONAL DESIGN CRITERIA

2.1. Available energy and load demand should be in balance as good as possible, nevertheless, with serious attention for the investment in photovoltaic modules and storage-batteries.
(Sheet II Load profile and available energy)
In fact there are two possibilities to design the system:

- A photovoltaic system simply in combination with a large storage capacity. Winter-shortness is covered by surplus during summer months. In this case it means unfortunately a battery-capacity of about 16000 KWh or an investment of 4 million dollars and a space of about 500 m^3.
- A photovoltaic system in combination with an aero-generator and a storage capacity of 200 KWh is far more preferable which can be seen easily after studying sheet II.

We decided to propose the latter one.

2.2. The available energy from photovoltaic solar system and aero-generator covers the schools energy-demand for 95%.
From the diagram II we learn an additional energy-surplus of about 30000 KWh net would occur during 8 months of the year.
Furthermore another 20000 KWh net can be used in summertime to feed the boardingschool related to the school of Maritime Studies. An extra cable 380/220 V ac of about 500 m length will connect this consumer to the system. This means, about 80% of the available energy will be used by two well defined consumers.
Considerations as mentioned above lead more or less automatically to the necessity of an Energy Management System.

2.3. Besides a major part of resistive load such as lighting, inductive loads (motors etc.) will be connected to the system.
(Sheet III Total System Approach)

3. THE ENERGY MANAGEMENT SYSTEM

Except the max. power point function and the application of a static inverter both with the aim of an increased system-efficiency, this proposal contains an Energy Management System too.

As can be learned from above, we deal with 3 direct energy sources:

- photovoltaic solar system
- aero-generator
- public utility

and one indirect energy source by a battery bank.

The EM system achieves both an energy flow as efficient as possible and guaranteed reserve by storage. A micro processor has been planned to control this, it will be programmed in a way that the expected available solar- and windenergy is divided in a suitable manner over the consumers and the accumulator-bank. The remainder will be delivered to the public grid.

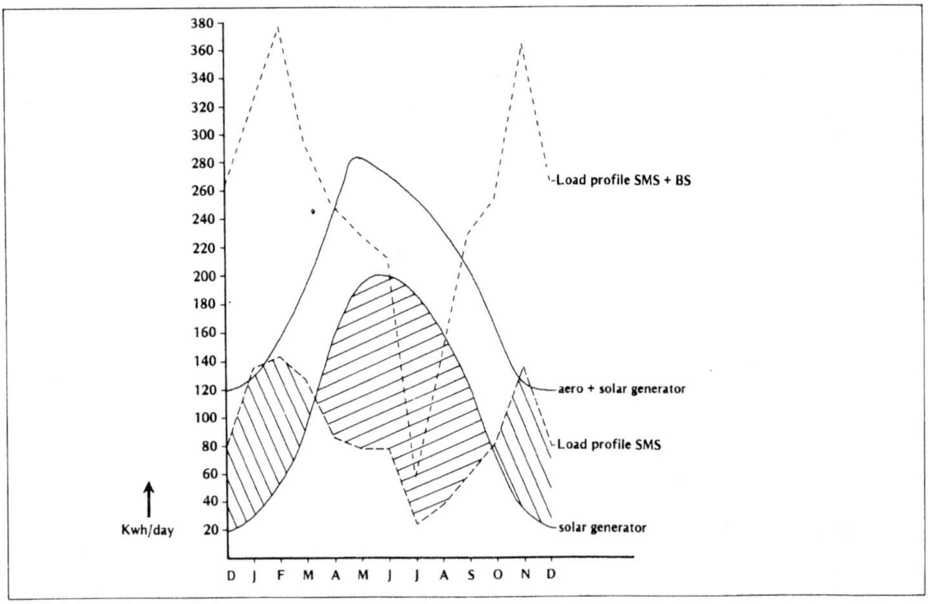

II. Load profile and available Energy

Solar System Terschelling

Members of the consortium:
HOLEC N.V., Holland
AEG - TELEFUNKEN, Germany

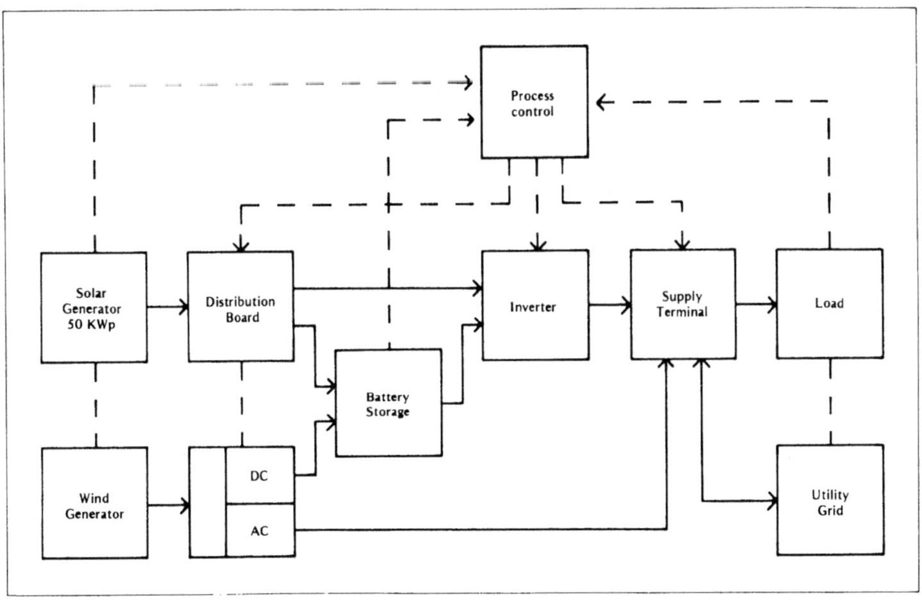

III. Total System Approach

Solar System Terschelling

Members of the consortium:
HOLEC N.V., Holland
AEG - TELEFUNKEN, Germany

ADRANO PHOTOVOLTAIC PROJECT

C. MALAGUTI and A. TASCHINI
Electric Research Center
ENEL-Milan

1. GENERAL

The aim of the Adrano PV Project is to test one flat-panels photovoltaic plant and another, photovoltaic concentrating plant suitable for use as parts of medium-sized and large solar power plants connected to the network.

The two plants will be located in Siciliy, close to the Eurelios 1 MW central receiver power plant; this will make it possible to compare the performance of the three different systems.

Each plant (flat-panels and concentrators) will consist of a 20 kW_p photovoltaic field, a maximum power tracker, and an inverter to feed the power generated into the local 50 Hz electric network.

The possibility of using a small battery storage will be considered in view of increasing the total efficiency of each plant, particularly in case of low intensity levels of radiation to be converted.

Wiring, conncectors, protection devices, and cable underground passages will be designed so as to combine high reliability with minimum cost.

The power conditioning subsystem (i.e., the maximum power tracker and the inverter) will be designed on the basis of its general compatibility with the photovoltaic systems, and will ensure about 85 % efficiency for power values between 25 and 100 % of the rated power.

All electrical, thermal and solar data effecting the operation of the two plants will be recorded by a dedicated data acquisition system, compatible and, therefore, able to operate connected with the supervision and data logging system used for Eurelios plant. In this way, the operation of the three plants may be compared in real time.

2. FLAT-PANEL PLANT

Silicon cells of at least four manufacturers of different European countries will be used for the flat-panel photovoltaic array.

Particular attention will be paid to the study of structures (construction, assembly using automatic equipment, maintenance), and the foundations, since, given the expected reduction in the cost of the cells, the structure supporting the panels would have an important bearing on the cost of the kWh produced by photovoltaic conversion.

In principle, all the 20 kW array will be supported by a single structure and all the panels will have a fixed orientation. This structure should be suitable, both from aesthetic and economic reasons, for use in the large solar power plants of the future.

This study will be conducted at ENEL, where research is already done on low-cost supports for electric power transmission lines and sub-stations.

3. CONCENTRATION PLANT

In the concentration plant, which is also having silicon cells, parabolic troughs from our different manufacturers will be compared. All systems will use a two axes tracking device; while the concentration ratio will be between 20 and 40 suns.

The type of cooling (natural or forced, air or cooling fluid) will be determined and optimized on the basis of the values of the previous variables.

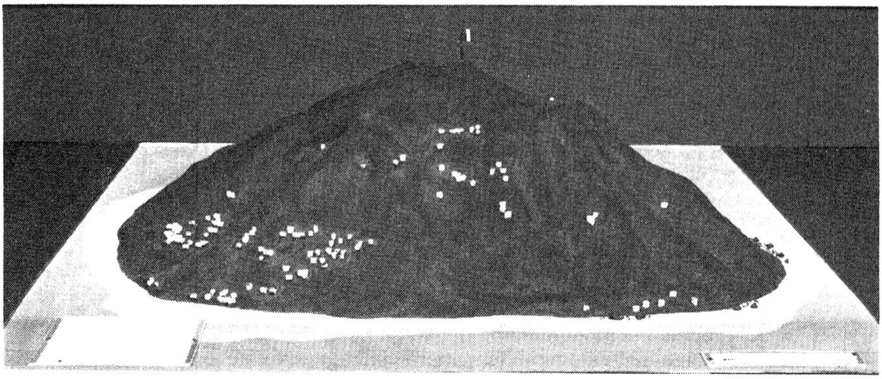

ALICUDI PROJECT

C. Malaguti and A. Taschini
Electrical Research Center
ENEL-Milan

1. GENERAL

The purpose of the Alicudi Project is to test the possibility of supplying electricity to a small community living in a remote site using locally-available renewable energy sources.

Particular attention will be given to the reliability, durability, efficiency, and ease of maintenance of the whole system. In fact, normal maintenance will have to be done by unskilled local staff.

The part of the Project, dealt with hereafter, consists in supplying, by means of a flat-panel photovoltaic plant, electric energy for lightning purposes, food conservation (refrigeration), and mass media (television).

The sizing of the flat-panel photovoltaic plant is based on the insolation available during the winter months. This means that more power will be available during the summer for purposes connected with tourism, which, as is known, is concentrated during the sunniest months.

It will be possible to make more energy available to the community in the course of a latter part of the Project (not considered at this stage) by installing one or more wind-driven generators with an individual capacity of about 50 kW, suitable for use in remote places. These generators are at present being studied and tested at ENEL.

2. THE SITE

For this experiment, a site was chosen - the Island of Alicudi - whose technical, environmental, and socio-economic characteristics were such as to be fully representative of localities where the electric power produced by photovoltaic conversion would, even at an early stage, be competitive.

Alicudi is the smallest and most outlying of the islands in the Aeolian Archipelago, which is situated north of Sicily, in the central Mediterranean.

The archipelago consists, for the most part, of islands of volcanic origin - two of the islands have volcanos that are still active - and the sea around them is very deep and often rough, which makes communication between the islands difficult.

Up to the end of the last century, the island of Alicudi was inhabited by about 800 people, most of whom were engaged in fishing and agriculture. Evidence of the latter activity may still be seen in the terraces constructed on the steep slopes of island, which rise to a maximum height of about 600 m.

In recent decades, however, the island has seen the number of its inhabitants shrink considerably, despite the mild climate, due to its isolation, the lack of services, and the hardships of daily life.

Nowadays, it is inhabited by about 100 people, most of them old, whose activities are mainly connected with the summer tourist season. They live on the southerm and eastern slopes of the island, where there are a few clusters of houses, some of which, especially at the higher altitudes, are now uninhabited. The other parts of the island are inaccessible .

Mostly, these houses have only one store. Their roofs, which slope slightly, act as a surface for collecting rainwater, which is stored in tanks and is the only source of fresh water for the inhabitants. Water is generally drawn and distributed with the aid of buckets; very few houses have hand-pumps to draw water with, or even tanks with a gravity flow.

The island has no ports, but only a small pier, where little boats can tie up.

Lastly, as regards roads, the island is extremely primitive, and mules are the only usable means of transport.

The exact location of the photovoltaic plant will be chosen after a more detailed analysis of load distribution. However, there are various areas with a southern exposure and terraced ground that would be entirely suitable for the purpose.

Location of the site for the plant will also take into consideration the position of the wind-powered motors - probably they will be installed on the southern ridge of the island. In this connection, it is interesting to note that the wind and the sun, utilized to generate electric power, complement each other extremely well over the course of the year. Thus, wind-power production is at its maximum during the winter, while photovoltaic production is at its height in the summer.

3. DESCRIPTION OF THE SYSTEM

The system consists of the following sub-systems:

- Flat-panel photovoltaic field.
- Maximum power tracker and battery- charger
- Electrochemical storage.
- Stand-by diesel engine for emergency battery charging.
- DC/AC inverter (in case of AC distribution).

- Meteorological station.
- Data acquisition system.

The photovoltaic field will have a power of about 80 kW_p at AM1, so as to be able to supply residents, as already stated, with electric power even during the months of low insolation. In order to be able to obtain maximum power from the field in the course of the year, without making the system more complex, there will be three different positions for the panels (winter, summer, and spring or autumn). In addition, there will be a fourth position, in case it is desired to experiment with fixed inclination throughout the year. The inclination of the panels shall be set manually by one person. The photovoltaic field will be installed, as already mentioned, on a terraced site with south exposure; this will make it possible to reduce considerably the area occupied by the field.

The fact that the place is difficult to reach, means that transporting the various materials required for building the plant will be a laborious business. Structures, in particular, will therefore have to be carefully designed so that they are easy to transport (possibly by helicopter) and install.

The maximum power tracker/battery charger will convert the variable voltage and current output values from the field into the values suitable to the battery or the load. Overall maximum power tracker/battery-charger efficiency should exceed 90%. Intensity levels of radiation lower then 50 W/m^2 will be converted.

The battery will be designed to supply local requirements for about two days.

The power of the stand-by diesel engine will be as low as possible and in any case sufficient for emergency charging of the battery, should the state of charge drop below a pre-established level. The power of the diesel group will be therefore around 10 kW.

A special study will be conducted to decide on whether to distribute the power produced in the form of D.C. or A.C.

D.C. distribution, possibly using dedicated inverters for user appliances for which A.C. would give added efficiency (for example, lighting), would seem to be the best solution from a technical point of view. On the other hand, there is a certain amount of resistance to the use of D.C. by local consumers, due to the difficulty of obtaining appliances, not to speak of the higher cost and limited choice of such equipment.

In the case of A.C. distribution, the inverter will ensure better than 90% efficiency between 25% and 100% of the rated load. It will also be subjected to low losses at no load, and introduce a total harmonic distortion of less than 5%.

In any case the local distribution network will be designed and constructed by ENEL on the basis of optimization criteria for the whole system. Each household will be limited as to the maximum amount of power. Users will be advised on the choice of appliances to install, to make the best use of the power available.

SCOLAR PROJECT

The study of a photovoltaic system to identify the potential for solar electric generation in the UK

A. Dichler Lucas Energy Systems Limited, England

The Scolar Project has been planned as a major step to identify the potential for photovoltaic generation in Europe. It involves the design, development and installation of a 100 kW solar array at a working power station in Southern England. The site was selected carefully in conjunction with the UK's Central Electricity Generating Board to combine the requirements of a suitable load with the available sunshine levels.

A key part of the project will be to monitor system performance to establish what output can be achieved from this type of system and how this relates to the theoretical performance. It is intended that the installation may be moved once its operation has been proven, to a location where it can offer better economic advantages in fuel saving. This may be in the Northern Isles of Scotland and it is hoped that a small array can be mounted coincidently to establish in advance the anticipated output at the final site.

Located at the initial site is one of the main development laboratories of the Central Electricity Generating Board so regular detailed monitoring will be possible. This site should also facilitate the assessment of performance for different load types. One of the planned applications is to supply electricity to a fuel pumping station whose peak consumption is well matched to the array output. When this is not in operation the load will be distributed within the site providing a different load profile.

It is understood that the development of an efficient overall system is central to the project so this paper identifies only some of the particular aspects worthy of special attention. The project will form a vital role within the E.E.C.'s programme in bringing added credibility by operating in conjunction with traditional generating plant.

The current E.E.C. programme to commission a number of large scale photovoltaic demonstration projects, will if properly executed, strengthen European industry's ability to design, install and operate photovoltaic power plant. However, two major considerations must weigh as importantly as the "hands on" experience of such systems. That is the participants must view the venture as a commercial activity, not solely as a fund-raising exercise. Secondly, great care must be taken to ensure that the performance data collected is both relevant and adequate. Such systems would be useless as a development tool if raw data were not recorded and processed to show the correlation between predicted and actual performance and identify reasons for discrepancy.

Several stages leading to the identification of final monitoring and D.P. requirements can be envisaged:

FIRST STAGE
- definition of purpose. This must be clearly defined at an early stage from both the participant and 'users' points of view.
- design philosophy. Items from the definition of purpose, which together dictate the:-
- equipment requirement and system design and the performance prediction.
- definition of relevent monitoring, data aquisition and subsequent processing requirements.

SECOND STAGE
- identification of relevent safety standards and equipment practices with all participants.

- development brief for equipment manufacturers subject to discussion regarding suitable transducers or measurement techniques and interface requirements.
- discussion and formulation of D.P. requirements and output presentation format necessary to monitor and explain overall system, subsystem and constituent part behaviour, and allow eventual correlation with behaviour prediction.
- explanation of differences.

THIRD STAGE
- Implementation and commissioning.
- effect of design/equipment changes either forced or requested on philosophy and definition of purpose.

All stages are interactive and the monitor subsystem will be a crucial element requiring very careful definition, planning and implementation. There has to be a high level of confidence in the data collected and the processed output data must reflect this confidence.

It is clear that the power conditioning unit must also receive careful attention to maximise system efficiency. Likewise, the structures and interconnection of the modules within the array will represent new areas for development since systems of this size have not been constructed in Europe before.

These items and the other key parts - modules and batteries - will all receive careful consideration identifying the objectives and meeting them in the planned approach outlined above. The detail of how the final installation will be designed and produced is not specified fully in this paper nor the draft proposal. This will not be prejudged but taken as an integral part of the project.

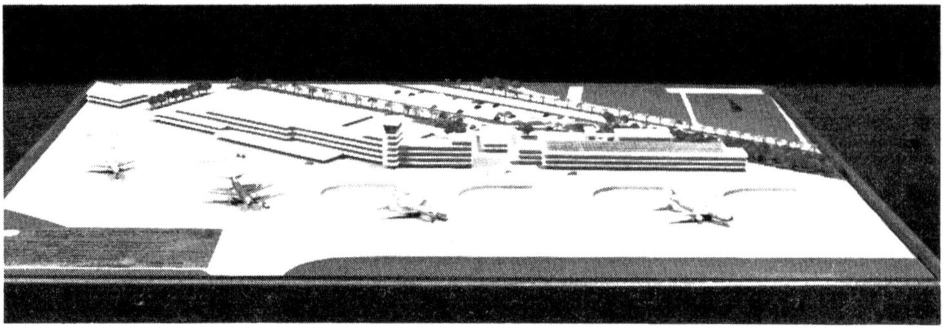

POWER SUPPLY OF ELECTRICAL ENERGY CONTROL SYSTEM OF AN AIRPORT

J. P. GOMIS — Executive Vice President

C. GRENIER — Operations Manager

of PHOTOWATT INTERNATIONAL SA.

ABSTRACT

This project proposes to use a photovoltaic power station installed on the roof of technical building of NICE Côte d'Azur airport to supply DC current to the electrical energy control system of the airport.
This system insures permanently the survey and the remote signaling of the electricity supply to each different consuming substation and controls the energy allocation throughout the airport.
The photovoltaic system with a peak power of 50 kW is the first example of the use of photovoltaic to perform an essential function on an airport. The system is sized in such way that it is entirely autonomous even during the winter months. The technical study is mainly directed toward the system reliability at all levels from the photovoltaic field itself to the charge control devices and the storage. The whole system is monitored permanently by a survey unit using a microprocessor to insure full reliability.

APPLICATION

For an airport such as NICE, the electrical energy distribution and control requires the gathering of all the main functions into a central control panel.
These functions are :

. remote control and signalling of each electrical station of the airport,
. remote control of emergency power station,
. control of energy allocation to the different stations,
. alarm network,
. display panels,

and other auxiliary services such as emergency lighting.
The electrical power needed is approximately 4kW.

For safety and reliability reasons those functions must be independant from the grid. A photovoltaic power station complies with this requirement giving to these functions a full autonomy from the main electrival distribution.

The proposed site is NICE, one of the most attractive resort of Europe, located in a sunny area. NICE airport ranks third among the french airports, more than 40% of its passenger traffic are related to international flights. The project will have wide visibility at the european level and will constitute an unique display for this new form of energy generation.

DESCRIPTION OF THE SYSTEM

The system will include the following subsystems :

- energy collection / peak power 50 kW
- energy storage of 115 kWh with related control and protection equipment
- user interface which includes an automatic commuting device to increase the reliability of the system
- control, monitoring and signaling unit for all the operating parameters.

Energy collection

It will be done with 720 PHOTOWATT modules installed on the roof of the technical building close from the main airport terminal and along the main landing strip.
The characteristics of PHOTOWATT modules are:

- circular monocrystal or semicrystal 100 x 100 mm cells,
- encapsulation laminated type (PVB or EVA),
- superstrate : tempered glass,
- substrate : aluminium / polyester composite,
- frame anodized aluminium,
- redondant cells to cell and string to terminals interconnections,
- junction box on back substrate,
- nominal power 35 or 70 watts,
- minimum power 30.5 or 61 watts,
- voltage at P max. 16 ± 1 volt or 32 ± 2 volts.

Solar array arrangement

To insure a high reliability level the photovoltaic array is divided in 45 parallel branches of 16 modules.
Each branch will be made of 4 parallel assembly of 4 modules connected in serie with redundant interconnection between the serie mounted modules.
The operating voltage of the array is 96 V and the modules tilt angle 60°.

Power conditionning

The two main functions of this subsystem are :
- control command
- storage

The CONTROL COMMAND unit will insure the charge regulation of the storage battery. It is designed in such way that the various branches of the photovoltaic array are monitored independantly. The charge regulation is of "cut off" type.

The STORAGE will be made of two independant strings 1200 Ah 48V each connected in serie when charge by the photovoltaic generator and used in discharge as two interchangeable 48 V batteries. This will insure to the storage unit the same reliability level than for the other components of the system.

Monitor and control unit

In order to provide the user with an efficient flexible and reliable control-command and monitoring system, the monitor and control unit(MCU) is a microprocessor based equipment.
The operation of the PV generator depends on following main factors :

. solar irradiance,
. temperature,
. load,
. availability of different sub-assemblies,
. ect..

An efficient monitor and control system will control all these factors, process the derived informations, compare some of them to a pre-programmed values and transmit orders to prevent dommage of certain important elements of the system.
On the other hand, this type of generators must be considered as a new feature of energy production : it is important to collect data for overall system analysis and later improvment of this type of systems.
Three main levels of functions could be identified for this equipment :

. level 1. detection and protection of catastrophic failures,
. level 2. mode select, operating command elaboration,
. level 3. statistics and analysis data base generation.

Some of level 1 functions are to be executed by a standard distributed equipment. Other functions will be executed by a centralized microprocessor based MCU.

System user interface

The DC network voltage is 48 volts. One battery string is used in normal operation, the second one being used as a back-up storage to insure permanent power availability and reliability.
For monitoring information, the switch from one battery block to the second one is fully automatic.

Security

To garanty to the user the highest reliability of the power system, a back-up exists from the grid. It consists of a batteries charger supplied by the grid and automatically controled by the monitoring system.

TECHNICAL INTEREST OF THE PROJET

Outside of the technical interest of the application by itself (insure essential functions of an airport) this project is an unique opportunity to study,design, install and operate a high reliability photovoltaic system. The experience acquired at each subsystem level will be of great interest for the futur photovoltaic power stations.

POWER SUPPLY FOR FM AND TV EMITTER STATION

C. GRENIER J. C. VIALART

Operations Manager - System Department Manager
of PHOTOWATT INTERNATIONAL SA.

ABSTRACT

This project experiments the use of 50 kWp photovoltaic system to supply frequency modulation emitters of TDF installed in the south of FRANCE. Emitters using AC power with a stable consumption the sizing of the solar system is done to satisfy the needs of one emitter during winter with an optimised volume of batteries and a small complement from the grid.
Due to this sizing the photovoltaic system will supply two (or more) emitters during summer.

APPLICATION

The user of this installation is " TELEDIFFUSION DE FRANCE ". This expérimental system will be installed at " Mont Bouquet " sunny area of south of FRANCE, near ALES.
The installations of " TELEDIFFUSION DE FRANCE " existing on this site are composed of :

. 3 frequency modulation emitters (France Musique, France Culture, France Inter)
. 3 television emitters one for each french station.

The FM emitters broadcast every day between 7 AP and midnight and the energetic needs of each emitter is stable during the whole working period: 6 kW for the 2434 emitter, or 102 kWh per day.
The fact to work mainly during the day is well adapted to the use of photovoltaic system as power supply.

These emitters are presently existing and supplied by AC from the grid. So to adapt our solar system to this existing materials we will convert the DC power from solar to AC according to the AC power supply specifications of the emitters.

But it is not impossible to forecast with the agreement of the emitter's manufacturer an other way to use DC solar power for a futur installation mainly if this installation has to be done in developping countries where AC grid is non existent in some areas.

This installation will be the absolutely necessary pilot experiment to qualify a technical solution which could be usefull in developping countries where exists a market for autonomous power supply for broadcasting.

DESCRIPTION OF THE PHOTOVOLTAIC SYSTEM

General presentation

The proposed photovoltaic power station is composed of :
. a solar array of 50 kWp power, providing energy collection,
. a 180 kWh electrochemical storage with protective devices,
. DC/AC conversion with two static inverters,
. automatic commutation of the load either on solar station or grid,
. data logging : to display and account the typical datas of the system.

Energy collection

Photovoltaic module manufactured by PHOTOWATT and COMPAGNIE FRANCAISE des PETROLES will be used for collection and convection of solar energy.
The general tentative data for PHOTOWATT module is :
. circular monocrystal or semicrystal 100 x 100 mm cells,
. encapsulation laminated type (PVB or EVA),
. superstrate : tempered glass,
. substrate : aluminium / polyester composite,
. frame anodized aluminium,
. redondant cells to cell and strig to terminal connectors,
. junction box on back substrate,
. nominal power 35 or 70 watts,
. minimum power 30.5 or 61 watts,
. voltage at P max. 16 ± 1 volt or 32 ± 2 volts.

The modules manufactured by CFP have the following characteristics : they are made by spraying the active material layers on a tempered glass 3 mm thick. The back of the module is made of severals layers of metallic deposit protected by a water proof seelant layer.The outside dimensions of the modules are 61 x 61 cm and their weight is 3.2 kg. Each module has an average power of 11.2 peak watt (25°C AM1).
The modules are assembled into arrays made of 8 modules. The arrays are framed with a specially treated wood (estimated life of the frame 40 years) The outside dimensions of the arrays are 137 x 260 cm and their weight is 45 kg each.
Electrical characteristics of one array :
. open circuit voltage .... 180 to 200 V
. short circuit current .... 750 vA
. voltage at max. power point .... 140 V
. current at max. power point .... 575 mA
. max. power .... 80 watts
The wiring between modules is integrated into the frame of the array. Each array will be equiped with two terminals.

Solar array arrangement

The photovoltaic field is composed of 60 parallel strings of 12 modules serial mounted, i.e. a total peak power of 50 kW with a nominal voltage of 300 V.
The tilt angle is 60°.

Power conditionning

This operation is based on simples and handy components providing to the station a good reliability.
Two mains functions are performed :

D1 . control of DC energy and storage
D2 . DC/AC conversion

D1. PRINCIPLE OF WORKING

The control unit for energy provides functions equivalent to those of the charge controllers of standard photovoltaic systems.
The storage batterie is connected to the solar array, the charge controller avoiding any overcharge of the battery as well as loosing of water.
The control is serial type, providing a limitation of the current from the array when the storage battery has reached a sufficient level of charge.
The storage is an electrochemical one. The total capacity is 180 kWh, by the way of 150 lead acid cells, low antimonious rate type with a self discharge lower than 5% per month.

D2. PRINCIPLE OF WORKING

Two static converters of 7,5 kVA supplies each chanel to fit at the best to the requirements. The wave shape delivered is well adopted to the utilisation and the control acts in order to avoid no-charge losses of the converters.

Safety functions

The status of charge of the battery is permanently checked. When a first level of abusive depth of discharge is reached, the second chanel, if provided, is automatically connected to the grid; at a second level, the load of the basic converter is automaticaly connected to the grid to protect the battery.

Data logging

This function shows the instantaneous status of the station by a data logging at various levels and also allows a general measurement by accounting of the different characteristics.
The interface is designed in order to keep the existing distribution structure. It allows to easily come back to existing energy distribution to avoid any trouble for the user.
This control of the grid connection of one or two chanels will be done by :

first ... the monitoring function of the system depending of the level of the batteries,
secondly .... following the daily duration during the year to avoid the multiple commutation between grid and solar.

THE FOTAVOLTAIC PROJECT : A 50 kWp PHOTOVOLTAIC PILOT PLANT PROPOSED FOR FOTA ISLAND, CORK, IRELAND

G.T. WRIXON
Microelectronics Research Center, University College, Cork, Ireland

Summary
It is proposed to build a 50 kW photovoltaic pilot plant on Fota Island near Cork. This is located in the South-East of Ireland, the region receiving the most solar radiation. The application is to supply energy to run a dairy farm. Because of the specific load requirements of the farm (peak load in summer), the system is well suited to cope with the yearly insolation profile. It is planned to use lead acid batteries to provide a one day storage only and thus the system will use the (already existing) utility grid as a back up. However, the energy balance is positive for each month. In order to meet the AC-requirements of the load, static invertors will be provided. To also ensure optimal efficiency in the partial load range, the power conditioning devices will be sub-divided into parallel inverter units. A specially designed microprocessor master control unit will supervise maximum power point tracking, battery charge/ discharge, inverter combinations, switching etc. as well as coordinate the taking of performance data.

1. INTRODUCTION

In thinking of a rational photovoltaic pilot plant for northern latitudes, it is desirable that ideally the following criteria be satisfied:

- the load should have a seasonal variation similar to the available sunlight.
- the system should help to reduce peak load demand on the electricity grid.

A project which fulfills these criteria and in addition is of economic importance in the Irish context is the application of photovoltaics to the dairy industry. In the south of Ireland sixteen times more milk is produced during the summer than during winter. Thus the total dairy industry (milking machines and parlors, creameries, dried milk, butter, cheese factories etc) has the correct seasonal variation. Consequently, accumulation of experience with a photovoltaic system applied to some aspect of this industry would be directly applicable to a realistic potential use of solar cells in Ireland at a time in the future when their use will be economically justified.

Fota Island is situated in Cork Harbour and has an area of about 200 Hectares. It has a global solar radiation of 1100 $kWhm^{-2}$ per year on a horizontal surface with a range of 155 $kWhm^{-2}$ in June to 16 $kWhm^{-2}$ in December.

Fota Island was purchased by University College Cork some years ago and is now the site of the University Farm which is administered by the University's department of Agriculture. Amongst other activities the Farm maintains a 250 head dairying herd. Currently the farm's peak electricity demand is about 11 kW being made up of milking machines, bulk refrigeration tanks, compressors, sterilizers, pumps and lights. It is proposed to use the photovoltaic generating station to satisfy the electricity requirements of this farm.

2. SOLAR GENERATOR AND STORAGE

The solar generator will be made up of flat silicon solar cell modules. The cells will have a size of 10 x $10cm^2$ and will be of multicrystalline silicon. According to the cell design progress by the beginning of the production phase the doping process, the contact material and the interconnection technique will differ from today's state-of-the-art. The cells will be glass-encapsulated and the module size will be similar to present-day module sizes around 0, 13 m^2 manufactured by AEG-Telefunken.

The generator will be of modular construction, the modules will be interconnected into strings. The single groups of strings will be switched onto the DC busbar. The preliminary design of the generator assumes an angle of tilt of 70°, which is due to the bad winter insolation and eases the yearly variation of the monthly average daily energy to 1:5.3.

Battery use and storage is required for the system for the following reasons:

- load equalization (daily); stand alone capability
- constant voltage in this case 220 V DC
- surge current source
- energy back-up in bad weather periods

The lead acid battery capacity chosen (700 Ah 220 V DC) represents a one day energy reserve for the dairy farm load requirement in bad weather periods or grid break down.

3. POWER CONDITIONING AND PROCESS CONTROL.

The power conditioning will consist of a number of static inverters connected in parallel and microprocessor controlled so that the individual inverters can be switched in or out in accordance with the momentary requirements. They will also be available to switch surplus energy into the grid when it is available or to take energy from the grid when the farm is not self-supporting. The microprocessor controller will oversee the switching of the inverters and that of the total system. The emphasis will be on self-sufficient operation so as to use the grid as little as possible. This is important, as periods of maximum power utilization (early morning and late afternoon) will not coincide with maximum input solar power.

PHOTOVOLTAIC POWER PLANT ZAMBELLI (VERONA)

G. DIETRICH, W. DÖNITZ, H.-J. HENSELER
DORNIER SYSTEM GMBH
Friedrichshafen

Summary

For the growing demand of potable water due to the development of tourism in a sunny area north of Verona a photovoltaic power plant of 140 kW (reduction to 70 kW minimum is possible) is proposed to supply the water station "Zambelli" with electricity. At the same time the electrical energy generated can be fed into the public grid. The peak demand of water during the summer coincides with high solar insolation. The flexibility of the proposed system is particularly suitable for demonstrating the utilization of solar energy and its use within the scope of public energy utility.

Special features and advantages of the planned system are:

- Absence of energy losses during storage of water
- effective utilization of solar energy using high efficient three phase inverters
- one-axis tracking of the solar cells to increase the total useful insolation and to improve the daily energy output of the electric system.

1. INTRODUCTION

In the Verona district, the population is supplied with potable water, natural gas and electrical energy by the local utility "Azienda Generale Servizi Muncipalizzati (AGSM). For this attractive region with a fast growing potential for tourism a further increase of the water supply capacity is particularly necessary at the "Zambelli" water station. For this station four water pumps are being planned. One of them (55 kW) will operate 24 hours/day while the remaining three pumps (20, 35, 42 kW) will be used for peak loads only.

Since the region to be supplied is touristically most frequented during the summer months, it is advantageous that peak water demand and peak insolation coincide. Since the area around the water station Zambelli is expected to yield 20 % more insolation than the hazy Verona valley it seems particularly suitable for demonstrating utilization of solar energy.

In addition, coupling of the photovoltaic power plant at this site with the existing electrical network poses no problems. This is considered by the energy utilities to be of importance if solar energy is to play a part in supplying European countries with electricity.

2. ELECTRICAL SYSTEM

The main design objective of the electrical system has been to obtain as high a share of the available solar energy as possible in form of useful electrical energy at the main coupling switch, without exceeding a reasonable investment cost level in relation to the installed capacity.

These requirements will be met by two inverters being suitable for operating synchronously with the grid, but also for selfcommuted operation. The last point is important for an efficient operation of the plant without grid connection. The inverters reach an efficiency of 90 % at the rated output, including filter losses, and deliver a voltage with negligible harmonic contents. In order to avoid a large reduction of the efficiency during partial load operation the power ratio of the two inverters will be choosen in the range of about 1 : 3. In addition, this measure allows to feed power into the public grid if there is more than 50 W/m^2 insolation at the solar cell panels. The auxiliary power supply for the control and regulation electronics is supplied by the dc solar generator, via a small dc-dc converter. This enables the operation of the plant independently of an auxiliary power supply from the grid, but it requires a carefully adjusted control for switching the plant on or off during all possible operating conditions.

3. LOAD CARRYING STRUCTURE

A further improvement of the system efficiency will be achieved by the use of a non-concentrating single axis tracking system with the axis of rotation parallel to the earth's axis. An energy gain of approx. 25 % can be expected by the tracking system in comparison with rigidly installed solar cells. In addition, the daily energy output of the electrical converter system will be advantageous with respect to total efficiency. Furthermore, the operation behaviour of the photovoltaic power plant will be improved considerably in particular for all systems without electrical energy storage.

The higher costs for the tracking system is justified by the significantly increased total energy output.

A cost estimation has shown, that the tracking system will be approximately 25 % more expensive than the structure of rigidly installed solar panels. These increased costs can easily be compensated, since the number of the solar cell panels can be reduced by 20 to 25 %. On the other hand the cost of the tracking system is only about 1/5 of the cost for the solar cell panels

Construction and operation of the whole photovoltaic plant will be done in close cooperation with the local utility (AGSM) and the joint solar energy research center CRAIES.

"NORTH EUROPE HYDRO PHOTOVOLTAIC PEAK POWER PRODUCING PLANT"

G.R.SMEKENS.

E.N.E. Energies Nouvelles et Environnement.
Brussels - Belgium.

Summary

The project involves parallel operation on the grid of a photovoltaic power generator with a run-of-river hydro-electric generator located on the same site. The purpose is to demontrate that through common management of both generators it is feasible to obtain north of the Alps a reliable peak power producing unit notwithstanding the random and seasonally varying nature of each individual energy generator. The precise goal to be achieved in this respect by the total system is to make available to the grid at the time when there is a demand for peak power, a guaranteed daily energy equal to 50% of the total yearly energy production of both generators and evenly spread 365 days per year. By making use of minimal storage requirements available behind the dam and in a small battery, the reliability of such a level of guaranted daily energy is to be 97%.

1. DESIGN BASIS

Within the EEC 1MW photovoltaic programme a combined hydraulic and photovoltaic peak power producing plant is proposed by a consortium by two Luxemburg and two Belgian companies on the site of an existing hydroelectric run-of-river plant on the Suur river in the Great Duchy of Luxemburg. (Slide 1.)

Such a combination of hydroelectric and photovoltaic energy in northern Europe will provide a seasonally levelled out energy supply throughout the year.(Slide 2 shows what can be achieved by a full scale plant on the Suur river). In addition to this, the random nature of the sum of both energies is reduced in comparison with the random nature of each individual component

These conclusions were drawn from a theoretical study and a simulation study carried out by an earlier EEC contract.

The pilot plant implementation of these conclusions will involve the coupling on the same bus bars of an existing 38kVA hydroelectric generator at Erpeldange, Great Duchy of Luxemburg with a 42kW peak photovoltaic panel installed at the same site for the convenience of common management. (Slide 3).

When applying conclusions drawn in the afore mentioned study, the total pilot plant system should be capable of producing a guaranteed daily energy of 220kWh/day, 365 days per year (50% of total yearly hydroelectric+photovoltaic net energy productions after storage losses) with 97% reliability. This goal assumes that on one hand there be approximately 40kWh net electrical energy stored in fast discharging batteries and that on the other hand there be approximately 300kWh net electric energy stored behind the dam which is the case at Erpeldange today.

The target guaranteed daily energy of 220kWh/day produced by the system will be obtained through operation of one alone or both types of generators simultaneously.

Considering the inverter efficiency the net photovoltaic power to the grid will be 38,6kW which could be added to the hydro-generator to totalise 76,6kW developped 2,86 hours/day. More frequently however, the two separate generators will run during different lengths of time but should totalise 220kWh's/day every day (except for 3% run out or 11 days downtime outside of the maintenance period).

Production planning with the Luxemburg grid will obviously attempt to maximise supply of the grid when there is a demand for peak energy, but it will also attempt to by pass the battery storage system to minimise storage losses as much as possible.

2. ARRAY DESCRIPTION (Slide 4.)

The array totalising 42kWatt peak will comprise 38 subassemblies of 40 modules each in the case of 3" squared monocristalline cells and 32 modules each in the case of 10cm by 10cm square cells made out of Semix polycristalline material. It is anticipated to have strings of 560 cells in series to acheive a nominal 220Volts. Parallel connections are foreseen. In accordance with the past fundamental study horizontal reflecting boosters will be provided. Their cost is expected to be 5 to 6 U.S.dollars per square meter of reflector area and at this price the extra kWh's supplied will have a cost as low as todays cost of peak energy (10 to 12 cents U.S.dollars/kWh). To achieve this, the tilt angle of the panels will be about 60° over the horizontal to maximise the effectiveness of the hydro-

SLIDE 1.

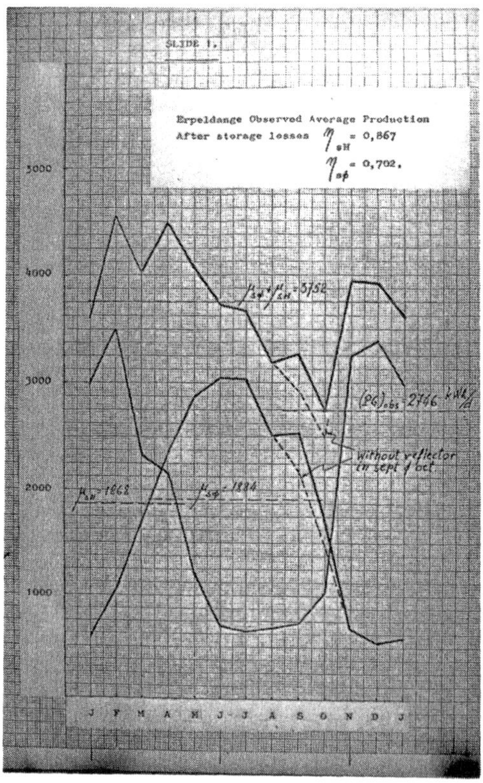

SLIDE 2.

SLIDE 3.

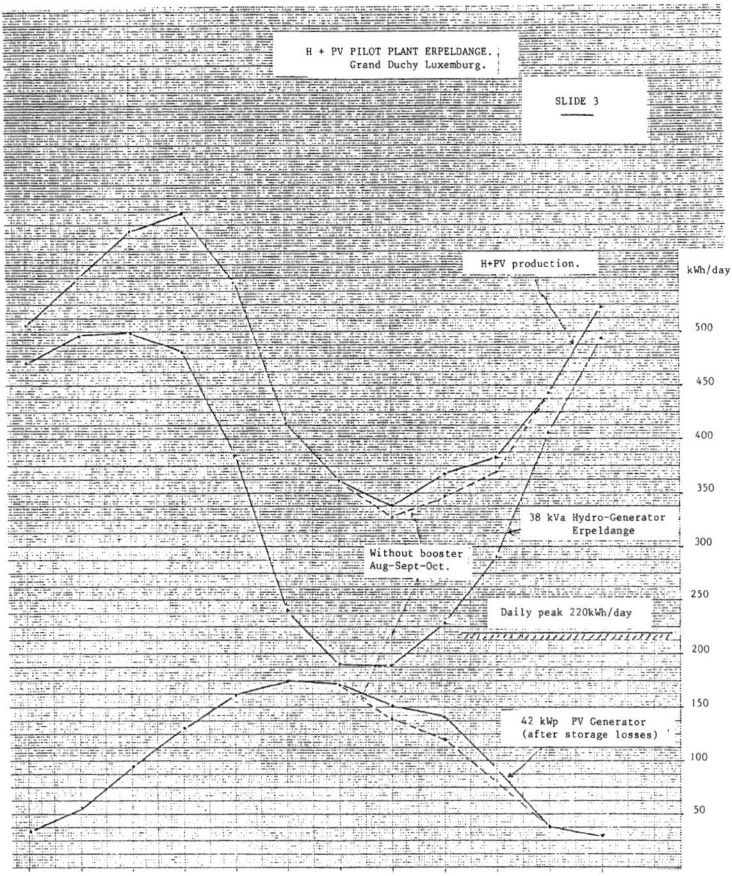

SLIDE 4.

photovoltaic combination, by raising its minimum yearly and thus increasing the peak power guaranteable every day of the year.

A supporting structure of the sub- ssemblies made out of wood appears to be the cheapest alternative at least for major supporting beams.

3. POWER CONDITIONNING

The fast discharging battery foreseen will be capable of delivering 40kWh's in one hour to the grid via the inverter.

The inverter will be designed on the basis of 92% efficiency at nominal power and with a total harmonics content on The A.C side not exceeding 3%. There must be no harmonic 5 at all to avoid interference with the grid remote control frequencies. Other features of the inverter will be automatic triggering of the discharge into the grid upon high voltage at the batteries and automatic cut-off and alarm in case of battery low voltage. The Luxemburg power distribution company intends to analyse the availability of the peak power generated after acheiving the project and will investigate the feasibility of remote control of the unit from a central dispatching station.

It does not appear economically justified to provide a maximum power tracking system.

Finally, evaluation of a power factor adjustement device will be made in cooperation with the Luxemburg grid operator, comparing advantages of a capacitive power factor with a possible loss in inverter efficiency.

REFERENCES.

1. 1979 Photovoltaic Solar Energy Conference, 23-26 April 1979 - Berlin (West). : "Study of a Hydro-Photovoltaic Plant for Peak Power Generation in Central and Northern European Countries" by G.R.SMEKENS, G.CARBONELLE and R.A.GAASCH, E.N.E. Brussels Belgium.

2. EC SOLAR ENERGY PROGRAM - Project C. Contrat N° 472-78-1ESB with E.N.E. Brussels- Final Report. June 1979.

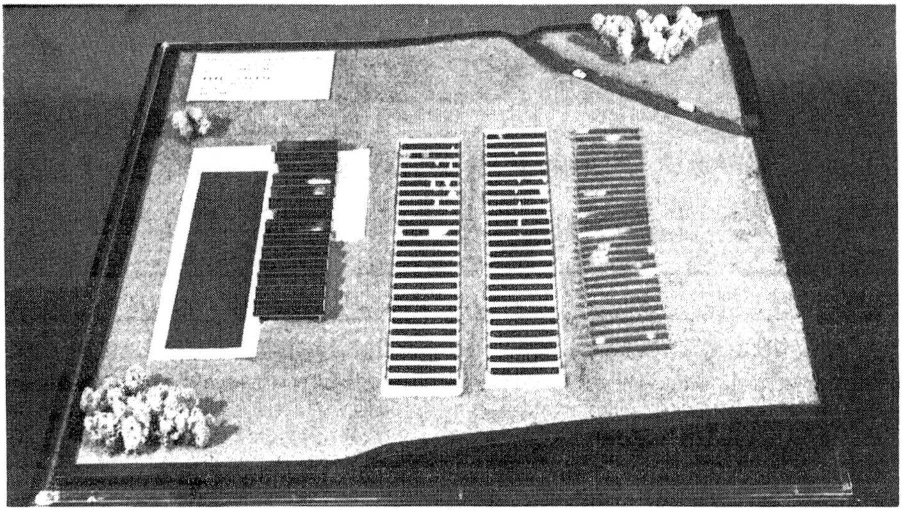

PHOTOVOLTAIC UTILITY CONNECTED TO THE GRID, FEEDING THE AUXILIARIES OF A SOLAR THERMAL SYSTEM

M. VAN GYSEL
R and D Manager
I.D.E. SA, Zoning Industriel, B - 5430 - Belgium

1. APPLICATION - ADAPTATION TO THE NEED

A major feature of the proposed project is the adaptation with respect to time, of the energy delivered by the photovoltaic system on one hand and the energy demand on the other.

The demand (some 25 kW in AC current) will come from the auxiliaries of the 2100 m^2 solar thermal system of Chevetogne, Belgium, and is thus in phase with the supply. This adaptation allows the system to be autonomous with a relatively small amount of battery storage. Alternatively, the system will be able to be connected to the grid instead of depending on its batteries.

2. MODULES

The module encapsulation and interconnection will not be a development item in this project. On the other hand, development in the fields of

- mechanical integration
- outside interconnects
- sorting of cells and modules for optimizing series - parallel arrangement

will be emphasized in view of integration in a large solar field.

3. MECHANICAL STRUCTURE OF THE PANELS

The mechanical structure supporting the panels will be addressed towards drastical cost reduction, in materials and installation labour.
Low cost materials such as galvanized steel, simplified structure (optimized in conjunction with the structure of the modules themselves), junction to the ground using poles will among others be studied for reaching this aim. The use of hedges or fences for minimizing wind load is envisaged.

4. INTERCONNECTIONS

Medium voltage (above 200 V) will be used in order to decrease line losses. Hot spot and safety problems will be addressed as a result of this. Monitoring of defects, repair under operation, lightning protection will also be allowed for.

5. INVERTER AND MONITORING

The inverter will be of the type developed with EEC financing in the project "A power conditioning interface for a photovoltaic utility".
It is a stored waveform inverter, microprocessor controlled, with a single chopper, achieving a typical efficiency of 95 %. The inverter is assembled through paralleling of 4 unit inverter modules of 10 kW.
It will be able to run the system in two configurations :

a. The basic configuration will use a small size battery storage, the inverter being able to manage charge and discharge of the batteries.
b. The same hardware will allow an alternative configuration to be tried, in which the system uses the grid as a buffer, without battery.
   In that case, the inverter will be piloted through a maximum power point circuit, driving the photovoltaic for maintaining maximum output.

Finally, the inverters will be coupled to the grid through a single 5 - leg transformer and an appropriate filter.

AGHIA ROUMELI - ISLAND OF CRETA, A 50 kW AUTONOMOUS PHOTOVOLTAIC POWER

C.N. KOUROGENIS (YEET), B. AUBERT (SERI Renault)
L. SELLES (SERI Renault), J. CHADJIVASSILIADIS (PPC)

1. INTRODUCTION

This project is funded and/or managed by the :

- Scientific Research and Technology Agency (YEET) of Greece
- The Commissariat à l'Energie Solaire of France
- and the Hellenic Industrial Development Bank.

It is proposed by a European Consortium formed by :

. SERI Renault Ingênierie of France

. VARTA A.G. of Germany

. The utility company of Greece called the Public Power Corporation

The project deals with an autonomous stand-alone 50 kW photovoltaïc power plant foreseen for a small village called Aghià Roumeli on the South Coast of the Island of Creta.

The plant will cover the energy needs of the village population.

The aim of the project is to demonstrate photovoltaïcs as an alternative to-day available to generate power to a remote and stand-alone community.

2. SITE DESCRIPTION

The village of Aghià Roumeli lies at the exit of the famous Canyon of Samaria overlooking the Mediterranean Sea.

As a matter of fact, the village consists of two hamlets ; one being located on the seashore, it's the new village ; the other being about 1500 m in the mainland pitched in the mountains, it's the old village.

The 34 houses of Aghià Roumeli are the dwelling of 105 permanent inhabitants. But it is a very famous touristic place in the Island of Creta.

The 200 beds of the hotels and the 8 shops as well as the campings in the vicinity are used to receive more than 170 000 tourists each year.

The 1520 modules will be supported by 63 rustic panels. These supporting panels will make use of metallic frames.

Each panel will support 24 modules connected in series.

. For the maximum power tracking device we are proposing a DC/DC converter that makes the PV array to operate at its maximum power point whatever is the battery voltage. This device would bring a gain of about 15 % on a year energy basis.

. Concerning the storage battery, we will mention that the selected elements from Varta are perfect for such an application : thanks to their low antimony/lead ratio, their self discharge remains in quite acceptable ranges at the operating temperatures concerned. Also, their large volume of electrolyte, in addition with their low evaporating rate are suitable to avoid frequent maintenance operations.

The installed capacity will be 500 kWh which corresponds roughly to a two day autonomy at full load.

. The 50 kW system will feature two 20 kVA inverters, mainly for a better reliability. They will be specially developed, using new transistors and the Pulse Width Modulation technology in order to reach efficiencies above 90 % at widely varying loads (let's say from 10 % to 100 % of the full load).

The wiring will be studied for a very fast cabling on the site. Like in our other project submitted for the Island of La Reunion, the input voltage of the inverters will be 300 V DC in order to increase their efficiency. They will deliver 220 V AC acting in phase when necessary to meet the loads requirements.

In conclusion, we would like to mention that this project will be carried out through a close cooperation between SERI Renault and the SCIENTIFIC RESEARCH AND TECHNOLOGY AGENCY OF GREECE.

To- day, energy needs are covered only seasonally by means of 7 Diesel gensets rated between 7 and 10 kVA. Fuel is supplied by the daily boat shuttle.

Efforts for the electrification of the village through the island's grid have been unsuccessful because the utility of the Island, the Public Power Corporation, is not allowed to draw a 15 km cable through the Samania Forest which is a National Park of an exceptional natural beauty.

On another hand, the coastal connection with the closest village has to be laid under the sea for a distance of more than 15 km.

So the only solution seems to be an autonomous power supply system that links the two parts of Aghiã Roumeli (the old and the new villages).

The Island of Creta is located close to the 35th parallel : in other words, to the same latitude of a major part of the North African Coast.

At Aghiã Roumeli, the yearly average irradiation is estimated to be 4.8 kWh a day.

However, because of its seashore climate, the outside temperature is not too high. It varies within the range of 25 and 35°C, during the summer.

3. TECHNICAL DESCRIPTION OF THE SYSTEM

The 50 kW PV generator will be installed in the vicinity of the new village. A pole line will be erected by the Greek Public Power Corporation to link the two villages.

The PV generator will consist of :

- the PV array field
- the DC DC converter
- the storage battery
- the inverters
- the central control command building

4. DESCRIPTION OF THE SUBSYSTEMS

. The 50 kW PV array will make use to a large extent of RTC BPX 47 C modules of 33 W each.

SOLAR PHOTOVOLTAIC COLD STORE

B. McNelis*

General Technology Systems Limited
Forge House, 20 Market Place,
Brentford, U.K.

1. INTRODUCTION

The use of solar energy for cooling applications is attractive for obvious reasons. Photovoltaic powered refrigeration plant can be effective and efficient, and will in the future be economically viable. This paper outlines a design concept for a photovoltaic agricultural cold store, a prototype of which is to be constructed in Sicily. The system has been developed from work completed by GTS (1, 2) and HELIX (3) to meet market requirements determined by AGIP.

The cold store will have a volume of approximately $750m^3$ and will utilise an array rated at 40kW peak. In the design the need for batteries for start up or energy storage is eliminated. The array drives compression cycle refrigeration plant directly and an ice bank is employed for energy storage.

2. DESIGN EVOLUTION

The problem common to all solar power systems is the mismatch between solar input and load demand. With photovoltaic systems the solution to this problem is in general use of a battery. This allows the varying output of the array to be stored and the load met as required. With a cooling system there is the obvious possibility of employing "cold" (e.g. chilled water or ice) for energy storage as this is considerably cheaper than using batteries. However, batteries are still required to run the compressor at its rated power and in particular to provide sufficient current to overcome start-up torque. A technical and economic analysis has shown that the most promising system configuration, which is also the simplest, is to use a variable speed d.c. motor driven directly by the array and a reciprocating compressor with cylinder unloading for load matching.

*representing consortium comprising, in addition to GTS:
HELIX Multi-professional Services,
Mortimer Hill, Berkshire, U.K.
AGRIMARKET SRL, Rome, Italy.

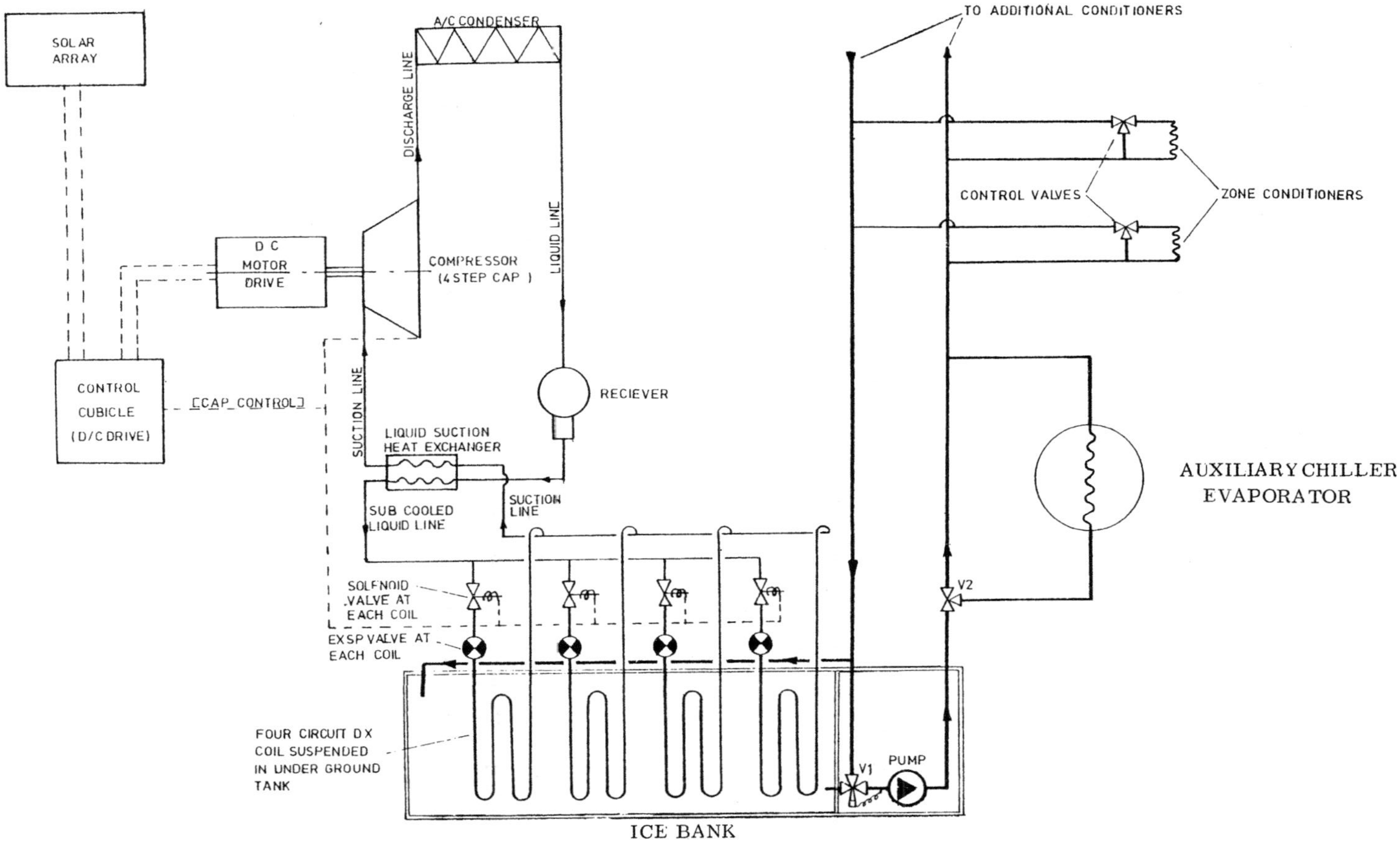

FIG. 1 - Photovoltaic cold store - system concept

3. SYSTEM CONCEPT

An outline of the system, at this stage based on a four cylinder compressor, is illustrated in fig. 1. The operating characteristics are such that the start is inhibited until the irradiance is sufficient to give an output of 70V on load. At this point with all cylinders unloaded the motor field coil is energised and a low voltage is applied to the armature. This is gradually increased, giving controlled acceleration of the motor and compressor, until the full output of the array is applied. At this point, one cylinder of the compressor is loaded. Motor speed and array output are monitored and these parameters used to load and unload cylinders to achieve very close matching, which is further improved by varying motor field current as a function of speed.

Direct expansion evaporator coils, which are also modulated in sequence with the compressor cylinders, are contained in a storage tank and are designed to enable ice to form directly on them. The ice bank provides both short and long term storage, and cold is transferred to the food storage space by chilled water/glycol. The system is designed for independent operation although in the pilot project a back-up chiller is included for monitoring and fail-safe purposes.

4. REFERENCES

1. B. McNelis, Proc. CEC Conf. "Solar Energy for Development", Varese (Italy) (1979).

2. W. H. Stephens, B. McNelis, A. I. Naylor, Proc. Photovoltaic Solar Energy Conf., Cannes (France) (1980).

3. C. Dodson, J. Keable, "Heat Pump Design for Building Research Establishment", (1977).

PHOTOVOLTAIC UNIT INTEGRATED WITH A HYDROGEN GENERATOR, PRESSURE STORAGE AND DISTRUBTION SYSTEM

P.M. SPAZIANTE, G. SIOLI and D.G. PHILIPPIDES

PANCLOR, SOLARIS, DORNIER Consortium

Summary

A pilot photovoltaic unit will be designed for 30 kW peak load and will be directly connected to the electrolysis unit generating hydrogen in unsteady state conditions. Low intensity levels will be converted into hydrogen at very high efficiency, due to the electrolyser characteristics. Hydrogen will be stored directly under 30 Bar pressure (no compressor needed). An industrial usage for H_2 will be provided (HBr production) and, alternatively, the H_2 generator will be reversible (fuel cell) producing electrical energy during the absence of the solar radiation. This energy well be locally used for process plant pumping and plant lighting.

The project

The project presented by the consortium formed by the companies PANCLOR, SOLARIS and DORNIER covers a pilot photovoltaic unit designed for a 30 kW peak load. The photovoltaic flat silicon panels will be directly wired to a water electrolyser operating in non-steady electrical load conditions and generating hydrogen at 30 Bar pressure. In order to obtain a uniform output hydrogen is stored under pressure,before being distributed.

In one of the embodiments hydrogen will be used for hydrogen bromide synthesis, while in a second one the hydrogen generator will be a reversible unit (eletrolyser and fuel cell) capable of generating electricity during the night.

The pilot plant will comprise the systems and sub-systems shown in fig. 1. The photovoltaic unit will supply direct current to the electrochemical unit, which performs an alkaline water electrolysis producing hydrogen and oxygen. Both gases are stored under pressure, without any need of mechanical compression, because the electrolyser is designed to directly operate at 30 Bar. The two gaseous products are then exported from the plant for usage. A utility system provides means for the thermal control of the electrolyser and of the storage unit.

A centralized control system will automatically set the operating factors. During the reverse operation (fuel cell mode), direct current will be exported from the plant.

Fig. 2 shows a plot plan of the installation. Two parallel structures, facing South, with a surface area of 400 square meters, carrying 1500 solar panels will be erected above a lagoon, within a marine salt-work. The wide unshaded site is ideal for the study of the contribution of strong light reflection of the aqueous surface. The lagoon is adjacent to a chemical factory producing marine water derivatives in Southern Italy. A model of the plant will be exhibited at the 1980 Photovoltaic Solar Energy Conference in Cannes. Fig. 3 shows an ideal coupling of the photovoltaic and electrolytic units. The voltage/current characteristics intercept where the power output is maximum at every sunlight level.

Furthermore, low intensity radiation will be converted to hydrogen at low current densities where the electrolyser efficiency is maximal.

The minimum voltage thermodynamically required to initiate the electrolytic process is guaranteed even under very low sunlight. Within these premises the system control is both easy and efficient.

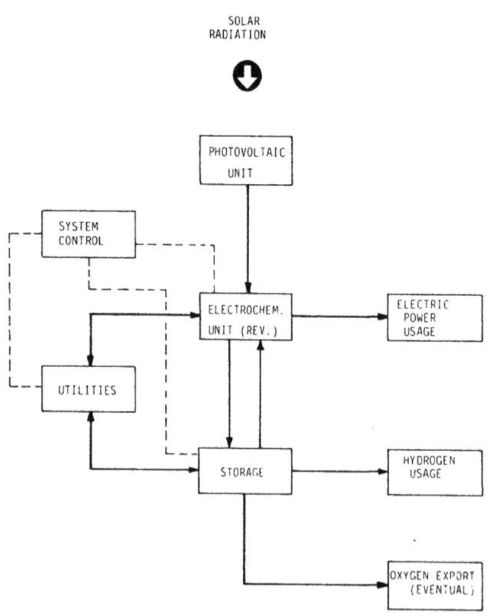

FIG. 1

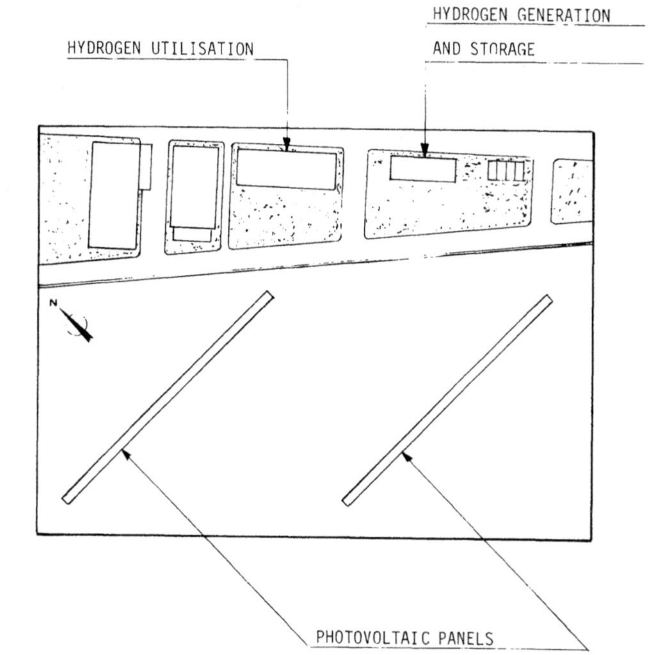
HYDROGEN GENERATION
AND STORAGE
HYDROGEN UTILISATION
PHOTOVOLTAIC PANELS

FIG. 2

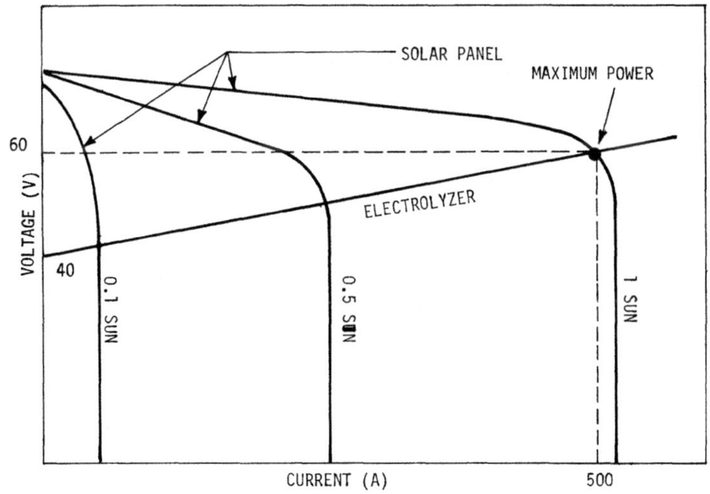
SOLAR PANEL
MAXIMUM POWER
60
40
VOLTAGE (V)
ELECTROLYZER
0.1 SUN
0.5 SUN
1 SUN
CURRENT (A)
500

FIG. 3

PHOTOVOLTAIC UNIT INTEGRATED WITH AN ACTIVE CHLORINE GENERATOR, STORAGE AND DISTRIBUTION SYSTEM

P.M. SPAZIANTE, G. SIOLI and D.G. PHILIPPIDES

PANCLOR, SOLARIS, DORNIER Consortium

Summary

A pilot photovoltaic unit will be designed for 30 kW peak load and will be directly connected to the electrolysis unit generating active chlorine in non-steady state conditions. Active chlorine will be directly stored as chlorinated water for later distribution.

Introduction

Photovoltaic silicon panels direct coupling to an electrolyser generating active chlorine is an efficient and "economic" system for water sterilisation. It is highly applicable to environments where both the requirements (sterilization and purification etc. of drinking water) and the starting materials (sun, salt) are present. One can recognize stereotypes of these environments, in towns near the sea, desert and mountainous regions or communities with water supplies in need of sterilisation.

The project

The project presented by the consortium formed by the companies PANCLOR, SOLARIS and DORNIER covers a pilot photovoltaic unit designed for a 30 kW peak load. The photovoltaic flat silicon panels will be directly wired to an electrolyser operating in non-steady electrical load conditions and generating active chlorine in dilute aqueous solutions by partial electrolysis of sodium chloride or directly from sea water.

This chlorinated water will be subsequently fed into the water supply system. The pilot plant will comprise the systems and sub-systems shown in Fig. 1. The photovoltaic unit will supply direct current to the electrochemical unit.

Figure 2 shows the plot of the current/voltage characteristics relative to the photocells and the electrolyser respectively.

A 30 kW peak-load from this installation will supply 6 kg/h active chlorine which evenly distributed into the water supply system is sufficient to sterilize 300 cubic meters of water per hour.

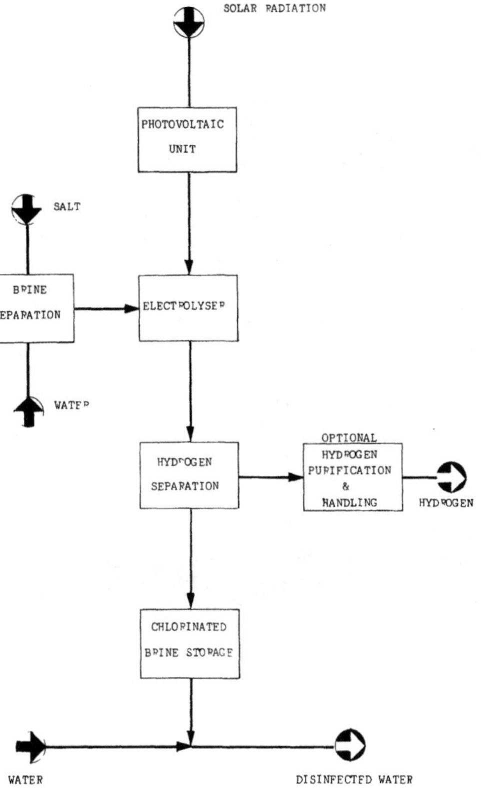
SOLAR RADIATION
PHOTOVOLTAIC UNIT
SALT
BRINE PREPARATION
ELECTROLYSER
WATER
HYDROGEN SEPARATION
OPTIONAL HYDROGEN PURIFICATION & HANDLING
HYDROGEN
CHLORINATED BRINE STORAGE
WATER
DISINFECTED WATER

FIG. 1

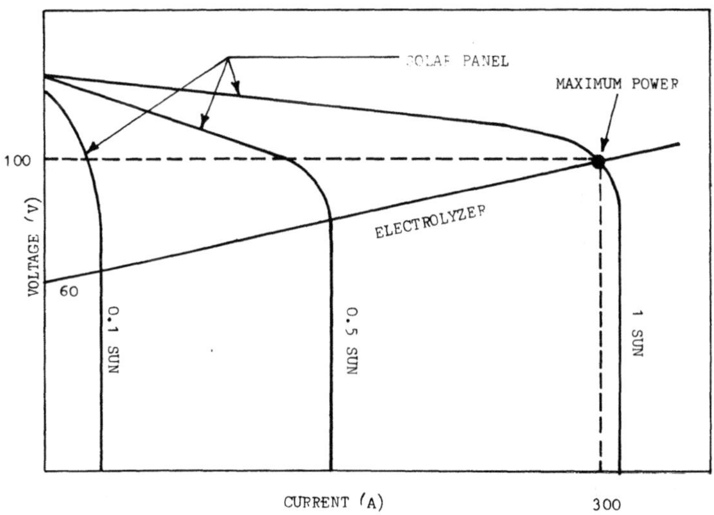
SOLAR PANEL
MAXIMUM POWER
100
VOLTAGE (V)
ELECTROLYZER
60
0.1 SUN
0.5 SUN
1 SUN
CURRENT (A)
300

FIG. 2

A 30 kW PHOTOVOLTAIC POWER GENERATING PLANT FOR INDUSTRIAL APPLICATION

Ph. M. KIRSCH

Energy Conservation Division, Engineering Manager

ACEC Belgium

SUMMARY.

The project is located in the area of Volaiville (Province of Luxembourg - Belgium). The site is characterized by a hill and a small river in a rural and forest zone of the country. The geographical properties have defined this site as a reference location for the installation of a multi alternative energy power station (Sun, wind, micro hydraulic and biomass). The proximity of a village permits also the interconnexion to an existing AC net.
The photovoltaic power plant is involving several improvements concerning the sun collectors, the electricity generation and the energy management.

1. SUN COLLECTORS.

In order to get accurate high voltage DC values individual cells having been classed by voltage at fixed intensity are used. Each cell is equipped with concentrator based on the prismatic principle. Radiators of the power electronic type are used. The supporting frame of the collectors are designed in order to integrate the various functions : supporting, cooling, concentrating and internal cabling.

The structure itself is made with light materials and composed by simple modules easy to handle and assemble. An economical optimization process has defined a low price collector system.

2. POWER GENERATION.

The DC current is transformed in a high efficiency inverter specially developped for solar application. Maximum power tracking electronic system is permitting to collect power even in relatively poor sunlight conditions. The electronic package is also assuring the interconnexion to external grids. (AC public net or other generator lines).

3. ENERGY MANAGEMENT.

A battery storage system is used in conjunction with microprocessor energy optimizers. These devices are integrating various parameters (Sunlighting, load demands...) to calculate and control the optimal use of the collected energy (Direct use or storage).

4. EFFICIENCIES.

The various improvements introduced at the different levels of the system lead to high efficiencies. Many losses are drastically limited (Dispersion, wiring, inverter...). The use of the prismatic concentrations will suppress the tracking problems occuring in some parts of the world, eliminating costly and power consuming auxiliaries.

The concentrators combined with cooling devices are also improving the economy of the system by preventing damages to the cells.

5. APPLICATIONS.

The use of such generation systems based on modular units can be imaginated in severe conditions for the developping countries basic electrification. (Isolated or interconnected). The use of electronics following space or military standards gives a full reliability to the system.
Life duration of 15 to 20 years can be expected with very low maintenance costs. Same if the to-day investment costs are still high for such plants, one must consider that no fuel expenses exist. Moreover, reductions of cell manufacturing costs must decrease considerably the investments of such installations in the next years.

80 kWp SOLAR PHOTOVOLTAIC PLANT FOR ISOLATED AREAS AND FOR INTEGRATION IN MULTISOURCES ENERGETIC ECONOMIES - DESIGNED BY COCSIER INT. CONSORTIUM ITALY (PROGR. 17 I+D)

R. VISENTIN
Director of CIRAES Università di Calabria Rende - Cosenza - Italy

Summary

The general characteristics of a 80 kWp solar photovoltaic power plant are analysed with reference to the local weather in the South of Italy at 39° North Latitude. Autonomous and not autonomous operation is considered for a proper choice of technologies and a proper exploitation of the local resources.

Introduction

The operation of the photovoltaic converter can be characterized by making reference to fig. 1 where S is the photov. converter, A the storage system, C the final channel of consumption of the energy, F a complementary source of energy. Let be E_M the maximum quantity of energy produced by the solar powerplant in a clean day with $\emptyset = .85$ KW/m^2 and no clouds in the sky from 8 a.m. to 4 p.m. and E_C the mean quantity of energy consumed per day in the final channel. Then $E_M/E_C > 1$ identify the "saturated systems" with direct storage A (Hydraulic, Hydrogen, mecchanical flyweels, electric batteries, compressed air etc.) used to accumulate the excess energy produced by S, while F is a complementary source of energy derived from the electric grid network or from local generators through other fuels. $E_M/E_C < 1$ identify the "not saturated systems", without the local storage A and integrated by F. In the case of "saturated autonomous systems" the totale quantity of energy produced in a year by S, exchanged with C and A, is equal to the annual consumption in C.

General characteristics of the 80 KWep solar powerplant

The solar powerplant will be made of n equal modules each one with 50 m^2 of flat photovoltaic panels of standard area 1x1,10 m^2, mounted on proper frames; 50% of the electric power will be produced by modules which will be allowed to track the sun being coupled to a 24 m long rotating axis, north-south oriented and with a 15° tilt angle over the horizontal plane.

The remaining 50% will be produced by fixed modules facing south with the same tilt angle. Each module will be placed inside a high transparent (90 ÷ 92%) metallic greenhouse and electrical power at 250 V d.c. will be exctracted by underground electrical connections. The 15° tilt angle is a typical slope of a large surface of the territory in south Italy as well as it is also the medium slope of the roofs of the houses.

Analysis of the plant operation

I Autonomous operation: let be E_o the mean electric energy demand per hour from the area C, E_1 the peak electric energy produced per hour by the solar powerplant in S when the maximum incoming solar flux is $\emptyset$=0,85 KW/m^2, η_s the energetic efficiency of the storage system A, n_s the equivalent number of hours of peak energy production per year, N = 8760 the number of hours per year, then for local autonomous production of electricity (1)

$$E_1/E_o = 1 + \frac{N - n_s}{n_s \ \eta_s} \qquad (1)$$

where in (1) the efficiency of the local electric network is placed = 1. From the data of fig. 2, by assuming η_s = .6 and with reference to the 15° tilt angle rotating plant, it follows $E_1/E_o \simeq 10$. Then the 80 KWep solar powerplant can supply $\simeq$ 8 KWe without interruption for all the year. Fig.3 shows the law of charge and disharge of the storage system A; hydropower

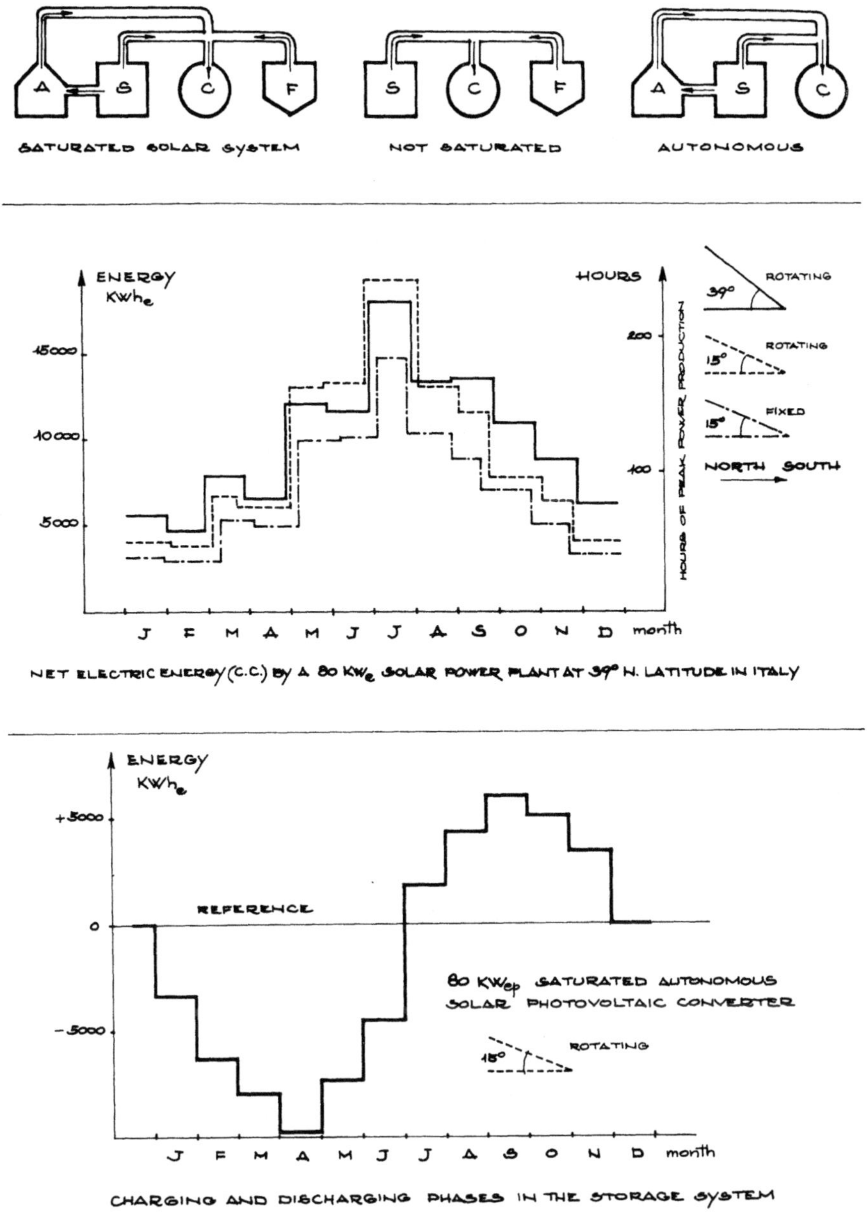

CO. C.S.I.E.R.
Proposal 17 I+D

FIG. 1

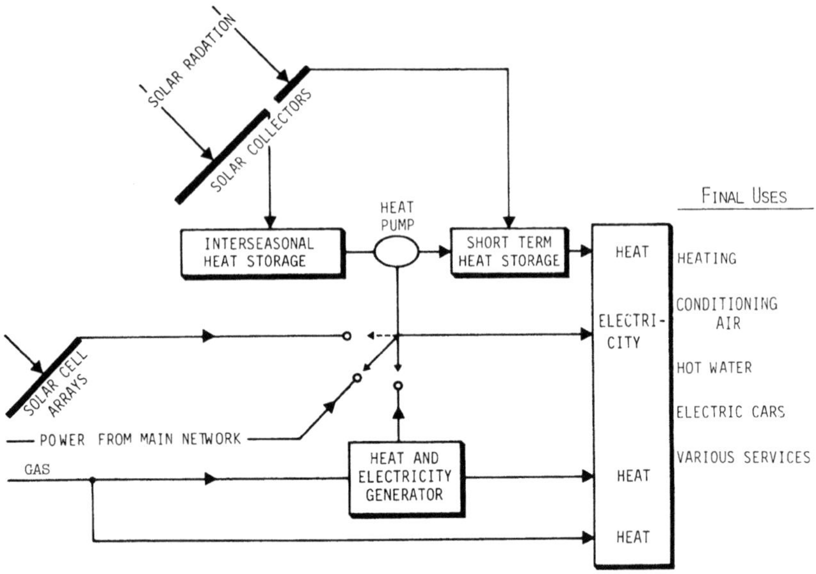

CO.C.S.I.E.R. SCHEMATIC OF THE SOLAR TOTAL ENERGY SYSTEM FOR THE BUILDING PROPOSAL 17 I + D

FIG. 2

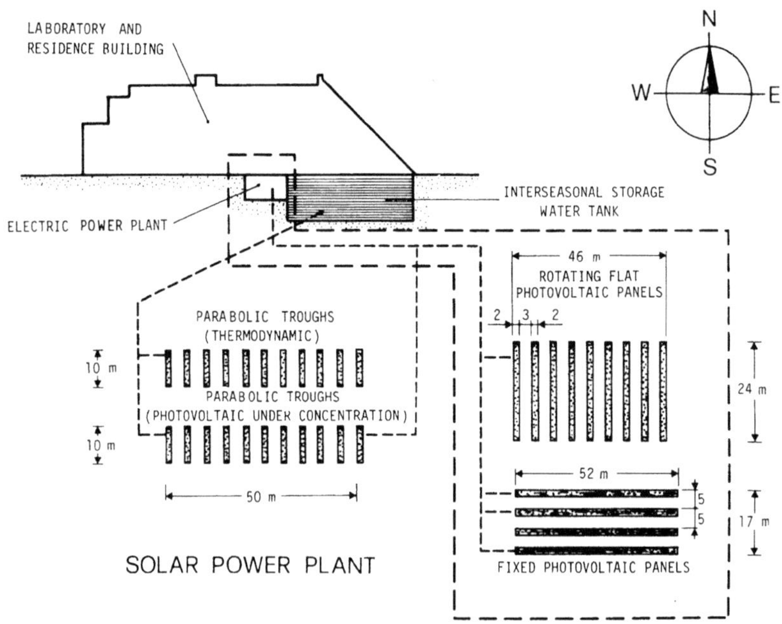

CO.C.S.I.E.R. CONNECTIONS SCHEME FROM SOLAR POWER PLANT TO ELECTRIC POWER PLANT AND STORAGE WATER TANK. PROPOSAL 17 I + D

FIG. 3

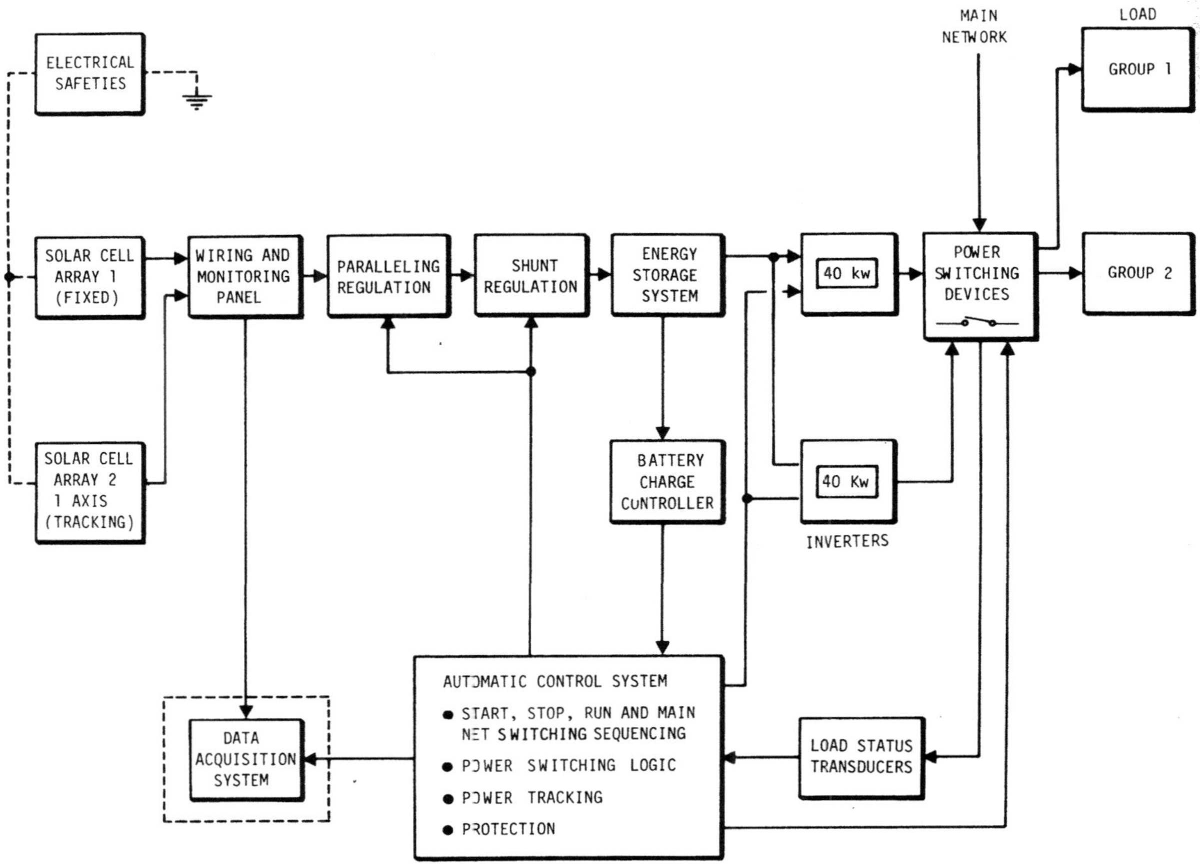

CO.C.S.I.E.R. PHOTOVOLTAIC SYSTEM BLOCK DIAGRAM PROPOSAL 17 I + D.

FIG. 4

storage as well hydrogen (electrolizer/hydrides/fuel cells) are good candidates for autonomous solar electricity production with storage efficiences η_S =.6. In this way 64% of the energy produced by the plant remains for the final channels of consumption: further energy losses from the conditioning subsystem (inverters, control apparata) with a reduction factor ≃ .75 have to be considered.

II Integration of the 80 KWep solar plant in multisources energetic economies

By inserting the plant in the energetic node of fig. 4 "indirect storage" of solar energy will be accomplished and the same performances described in fig. 4 will be obtained at higher efficiency; solar reduce the load in the electric grid network and/or in the natural gas network, then other fuels will be saved and they will be consumed to compensate the not continuous power generation from solar. An other approach is the following: the solar power plant will operate as an "energy saver" and it will save in a year the energy burned in a particular final channel in a shorter period. Our experiment on the solar assisted heat pump is an example of this approach.

The solar electricity will be inserted in the low voltage grid (220V) of a main final channel of consumption where the installed power is much greater than solar power (for example the 1 MWe of the Calabria University - Rende Cosenza (4)) and of which our particular final channel of energy consumption is a portion. Local storage of electricity with electric batteries is foreseen for 1/2 - 1 h of autonomy to compensate emergency effects due to a too reduced load from the main channel. Extensive use of electric batteries is foreseen only for applications to electric cars. This is a general approach for the proliferation of such "energy saver solar power plants" in south of Italy for civil and industrial applications: this proliferation is favoured by the properties of the territory (mountains, hills, roofs with slope facing south) and from the existing wide electric and gas network.

References

(1) R. Visentin — Proceedings of the Comples XVIII Int. Conference Vol. III pag. 254/267 Milano Sept. 79 Ed. ESA Via della Polveriera, 37 - 00187 Roma

(2) R. Visentin et alt. — CIRAES UNICAL Progress Report N° 2 January 1980 P.O.BOX379 Cosenza Italy

(3) R. Visentin et alt. — Energia Solare: Prospettive aperte dalle ricerche del CNR Ed. CNR - Peg. Via Fll. Bressan, 2 - MI

(4) COCSIER PROPOSAL TO EEC 17(I + D)

LIST OF PARTICIPANTS

ADDISS, R.

SOLAR POWER COPROATION
20 Cabot Road
WOBURN, Mass., 01801
U.S.A.

Tel. : (617) 935.4600
Telex : 710.348.0602

AUBERT, B.

SERI RENAULT INGENIERIE
2, av. du Vieil Etang
78.390 BOIS-D'ARCY
France

Tel. : (3) 043.81.34
Telex : 695.426 F

BARKATS, G.

AEROSPATIALE-FRANCE
B.P. 52
100, bd. du Midi
06.322 CANNES-LA-BOCCA

Tel. : (93) 479.600
Telex : AEROCAN 470.902 F

BEER, G.

PHOEBUS
Via G. Leopardi, 148
95.127 CATANIA
Italy

Tel. : (95) 38.29.45
38.69.01
Telex : 979.191 ENEL CT PER PHOEBUS

BERNDT, D.

VARTA RESEARCH AND DEVELOPMENT CENTER
Gundelhardtstr. 72
6233 KELKHEIM
Germany

Tel. : (6195) 802.312
Telex : 04.10.502

BERTELS, G.

ELECTRO-NAVALE ET INDUSTRIELLE
Kontichsesteenweg 25
2630 AARTSELAAR
Belgium

Tel. : (31) 874.081
Telex : 31.598 ENI B

BLOSS, W.

UNIVERSITY OF STUTTGART
Boeblinger Str. 70
7000 STUTTGART
Germany

Tel. : (711) 665.411

BRUNING, G.

SIEMENS AG
Werner-Von-Siemens-Str. 50
8520 ERLANGEN
Germany

Tel. : (9131) 77.220
Telex : 629.210

BURGESS, E.L.

SANDIA NATIONAL LABORATORIES
ALBUQUERQUE NM, 87185
U.S.A.

Tel. : (505) 844.2178

CAMPANELLI, V.

COMMISSION OF THE EUROPEAN COMMUNITIES
Directorate-General V for
Employment and Social Affairs
200, rue de la Loi
1049 BRUSSELS
Belgium

Tel. : (2) 736.60.00 - ext. 1032
Telex : 21.877 COMEU B

CANIPELLE, A.

LA RADIOTECHNIQUE-COMPELEC
130, av. Ledru-Rollin
75.540 PARIS CEDEX II
France

Tel. : (1) 355.44.99

CASPAR, R.

VARTA INDUSTRIE
165, rue J.P. Timbaud
92.400 COURBEVOIE
France

Tel. : (1) 789.01.54

CHAUVALLON, G.

AEROSPATIALE - FRANCE
100, bd. du Midi
B.P. 52
06.322 CANNES-LA-BOCCA
France

Tel. : (93) 47.96.00
Telex : AEROCAN 470.902 F

CHENON, J.	OMERA 49, rue F. Berthoud 95.101 ARGENTEUIL France Tel. : (3) 982.09.42 Telex : 696.797 F
CHEVALIER, Y.	COMMISSARIAT A L'ENERGIE SOLAIRE 208, rue Raymond Losserand Paris 75014 FRANCE Tel. : (1) 545.67.60 Telex : 230712 COMES F
CORDES, V.	AEG-TELEFUNKEN (N 461) Industriestr. 29 2000 WEDEL Germany Tel. : (4103) 702.833 Telex : 218.95.20
DELCROIX, J.C.	SERVICES DE PROGRAMMATION DE LA POLITIQUE SCIENTIFIQUE 8, rue de la Science 1040 BRUXELLES Belgium Tel. : (2) 230.41.00 Telex : PROSCIENT BRU B 24.501
DENIS, D.J.	OMERA 49, rue F. Berthoud 95.101 ARGENTEUIL France Tel. : (3) 982.09.42 Telex : 696.797 F
DICHLER, A.	LUCAS ENERGY SYSTEMS LTD. Windmill Road Haddenham Aylesbury BUCKS HP17 8JB Great Britain Tel. : (844) 291.681 Telex : 831.65 LSO HAD
DÖNITZ, W.	DORNIER SYSTEM GmbH Postfach 1360 7990 FRIEDRICHSHAFEN Germany Tel. : (7545) 82.847 Telex : 07.34.359

DOLLERY, A.A.

ROYAL AIRCRAFT ESTABLISHMENT
Space Dept.
Farnborough
HANTS GU14 GTD
Great Britain

Tel. : (252) 24461 ext. 2685
Telex : MODPE 825.332

DUMONT, A.

E.T.C.A.
B.P. 97 6000 CHARLEROI
Belgium

Tel. : (71) 366.525

DUMONT, Ph.

ACEC
STG/ENG 39.4
B.P. 4 6000 CHARLEROI
Belgium

Tel. : (71) 442.179
Telex : 51.227 ACEC B

DURAND, H.

COMMISSARIAT A L'ENERGIE SOLAIRE
208, rue Raymond Losserand
Paris 75014
FRANCE

Tel. : (1) 545.67.60
Telex : 230712 COMES F

FABRE, E.

LABORATOIRES D'ELECTRONIQUE ET DE PHYSIQUE APPLIQUEE
3, av. Descartes
94.450 LIMEIL-BREVANNES
France

Tel. : (1) 569.96.10
Telex : 203.475

FABRE, M.

MINISTERE DE LA COOPERATION (FRANCE)
Secrétariat d'Etat à la Recherche Scientifique et Technique
B.P. 3218
DAKAR
Senegal

Tel. : 21.79.33

FALCIANI, P.

OFFICINE GALILEO
Localita Tomerello
CAMPI BESENZIO (FIRENZE)
Italy

Tel. : (55) 4796
Telex : 570.126 GALILE I

FLECKENSTEIN, V.	SIEMENS AG Abt. E STE 13 Günther-Scharowsky-Str. 2 8520 ERLANGEN Germany Tel. : (9131) 772.75 Telex : 629.871
FONZI, F.	ADRIATICA COMPONENTI ELETTRONICI S.p.A. 67039 SULMONA (AQ) Italy Tel. : (864) 33.441 Telex : 60.049
FRIEDRICH, F.	KERNFORSCHUNGSANLAGE JÜLICH Projektleitung Energieforschung Postfach 1913 5170 JÜLICH Germany Tel. : (2461) 614.743 Telex : 833.556 KFA D
GAASCH, R.A.	INDUSTRIE DU BOIS S.A. 46, rue de l'Industrie DIEKIRCH Luxemburg Tel. : (352) 83.421
GAUTIER DI CONFIENGO, E.	S.E.S. Via Cuneo, 20 TORINO Italy Tel. : (11) 260.01 Telex : 22.10.50
GODDAER, M.	FABRICABLE 79, rue du Marché 1000 BRUXELLES Belgium Tel. : (2) 356.73.21
GOMIS, J.P.	PHOTOWATT INTERNATIONAL S.A. 98 ter, bd. Héloïse 95.102 ARGENTEUIL France Tel. : (1) 982.09.13 Telex : 698.850 F

GRASSI, G.

COMMISSION OF THE EUROPEAN COMMUNITIES
Directorate-General for Research,
Science and Education (DG XII)
200, rue de la Loi
1049 BRUSSELS
Belgium

Tel. : (2) 736.60.00 ext. 6801
Telex 21.877 COMEU B

GREBY, E.

DANISH ENERGY TECHNIQUE LTD.
Frederiksholms Kanal 16
1220 COPENHAGEN K
Denmark

Tel. : (1) 132.211
Telex : 63.481 UDSEN DK

GRENIER, C.

PHOTOWATT INTERNATIONAL S.A.
98 ter, bd. Héloïse
95.102 ARGENTEUIL
France

Tel. : (1) 982.09.13
Telex : 698.850 F

GUCHER, P.

INSTITUT DE CHIMIE ET PHYSIQUE INDUSTRIELLES
DE LYON
Service Electronique
31, pl. Bellecour
69.002 LYON
France

Tel. : (7) 837.52.86

HEINZELMANN, P.J.

BUNDESMINISTERIUM FÜR FORSCHUNG
UND TECHNOLOGIE
Heinemannstr. 2
5300 BONN 2
Germany

Tel. : (228) 593.542
Telex : 885.674 BMFT D

HENSELER, H.J.

DORNIER SYSTEM GmBH
Postfach 1360
7990 FRIEDRICHSHAFEN
Germany

Tel. : (7545) 83.766
Telex : 073.43.59

HIGGINS, A.

ELECTRICITY SUPPLY BOARD
27, Lower Fitzwilliam Street
DUBLIN 2
Ireland

Tel. : (1) 76.58.31
Telex : 30691 ESBC EI

IMAMURA, M.S.

MARTIN MARIETTA CORPORATION
B.O. Box 179
DENVER, Color. 80201
U.S.A.

Tel. : (303) 977.0701

JACQUIN, J.P.

POMPES GUIMARD
179, bd. Saint Denis
92.402 COURBEVOIE
France

Tel. : (1) 788.50.52
Telex : 610.592 F

KAESEN, K.

SIEMENS AG
Frankfurter Ring 152
8000 MÜNCHEN 46
Germany

Tel. : (89) 3500.899
Telex : 52.108.70

KEAVENY, D.

AEG-TELEFUNKEN
Industriestr. 29
2000 WEDEL
Germany

Tel. : (4103) 702.828
Telex : 218.95.20

KERN, E.C.

M.I.T. LINCOLN LABORATORY
P.O. Box 73
LEXINGTON, Ma. 02173
U.S.A.

Tel. : (617) 862.55.00 ext. 213
Telex : 923.355

KIPPERMAN, A.H.M.

EINDHOVEN UNIVERSITY OF TECHNOLOGY
P.O. Box 513
5600 MB EINDHOVEN
The Netherlands

Tel. : (40) 474.118
Telex : 51163 THENL

KLEINERT, K.	VARTA BATTERIE A.G. Dieckstr. 42 5800 HAGEN Germany Tel. : (2331) 393.397
KOEPKE, R.	KERNFORSCHUNGSANLAGE JÜLICH Projektleitung Energieforschung Postfach 1913 5170 JÜLICH Germany Tel. : (2461) 615.211 Telex : 833.556 KFA D
KORUPP, K.H.	AEG-TELEFUNKEN N461V2 Industriestr. 29 2000 WEDEL Germany Tel. : (4103) 702.846 Telex : 218.9520
KOUROGENIS, C.N.	SCIENTIFIC RESEARCH & TECHNOLOGY AGENCY 48, Vas Constantinou Ave. ATHENS T 501 Greece Tel. : (1) 742.589 Telex : 214.074 YEET GR
KREBS, K.H.	COMMISSION OF THE EUROPEAN COMMUNINITIES Joint Research Centre Ispra 21020 ISPRA (VARESE) Italy Tel. : (332) 780.131 Telex : 380.042 I
McNELIS, B.	GENERAL TECHNOLOGY SYSTEMS LTD. Forse House, Market Place BRENTFORD, MIDDLESEX TW8 8EQ Great Britain Tel. : (1) 568.58.71 Telex : 28604 (Ref. 599)

MACOMBER, H.L.	MONEGON LTD. 4, Professional Drive Suite 130 GAITHERSBURG, Maryland 20760 U.S.A. Tel. : (301) 258.75.40 Telex : 710.828.9617
MAGET, H.J.R.	VARIAN ASSOCIATES INC. 611, Hansen Way PALO ALTO, California 94022 U.S.A. Tel. : (415) 493.40.00
MALAGUTI, C.	ENTE NAZIONALE PER L'ENERGIA ELETTRICA Bastioni di Porta Volta, 10 20121 MILANO Italy Tel. : (2) 8847.8254 Telex : ENEL-MI 310.496
MALTAGLIATI, M.	SOLAREX V.C. Bini, 44 FIRENZE Italy Tel. : (55) 490.310 Telex : 570.126
MAYCOCK, P.D.	U.S. DEPARTMENT OF ENERGY Photovoltaic Energy Systems Division 600 E St. N.W. - MS 404 WASHINGTON, D.C. 20585 U.S.A. Tel. : (202) 376.19.50
MERCIER, D.	LEROY-SOMER Usine des Agriers 16015 ANGOULEME CEDEX France Tel. : (45) 624.111 Telex : 790.244 F
MERTIG, D.	AEG-TELEFUNKEN Industriestr. 29 2000 WEDEL Germany Tel. : (4103) 702.825 Telex : 218.9520

NOKERMAN, Ch.	E.N.I. NV. Postbus 389 2000 ANTWERPEN Belgium Tel. : (31) 874.081 Telex : 31598
PALZ, W.	COMMISSION OF THE EUROPEAN COMMUNITIES Directorate-General for Research, Science and Education (DG XII) 200, rue de la Loi 1049 BRUSSELS Belgium Tel. : (2) 736.60.00 ext. 6922 Telex : 21877 COMEU B
PETER, F.	COMPAGNIE FRANÇAISE DES PETROLES TOTAL 5, rue Michel-Ange 75781 PARIS CEDEX 16 France Tel. : (1) 524.46.46 Telex : 611992 TOTAL F
PHILIPPIDES, D.	PANCLOR S.p.A. Via Solferino 18/20 20019 SETIMO MILANESE Italy Tel. : (2) 328.51.41 Telex : 73428 CH
PIZZINI, S.	HELIOSIL Via Franchetti, 1/A 20100 MILANO Italy Tel. : (21) 24.70.15
RATAJCZAK, A.F.	NASA LEWIS RESEARCH CENTER MS 49-5 CLEVELAND, Ohio 44135 U.S.A. Tel. : (216) 433.40.00 ext. 6816 Telex : 985.218

RECHOU, J.	MOTEURS LEROY-SOMER Usine des Agriers 16015 ANGOULEME CEDEX France Tel. : (45) 624.111 Telex : 790.244 F
REMOVILLE, J.	LUXATOM p/o GRADEL 38, rue de Luxembourg STEINFORT Luxemburg Tel. : (352) 39.757 Telex : DEHA LU 2341
SCHNELL, W.	COMMISSION OF THE EUROPEAN COMMUNITIES Directorate-General for Research, Science and Education (DG XII) 200, rue de la Loi 1049 BRUSSELS Belgium Tel. : (2) 736.60.00 ext. 5697 Telex : 21877 COMEU B
SELLES, L.	SERI RENAULT INGENIERIE 2, av. du Viel Etang B.P. 19 78.390 BOIS-D'ARCY France Tel. : (3) 043.81.34 Telex : 695.426 F
SMEKENS, G.R.	ENERGIES NOUVELLES ET ENVIRONNEMENT 188, av. Van der Meerschen 1150 Bruxelles Belgium Tel. : (2) 771.13.28
SONNEVILLE, R.	HOLECSOL SYSTEMS B.V. P.O. Box 2300 5600 CH EINDHOVEN The Netherlands Tel. : (40) 524.655 Telex : 59.030 HOSOL NL

STAMMERS, J.R.	SOLAR TRADE ASSOCIATION LTD. 26, Store Street LONDON WC1E 7BT Great Britain Tel. : (1) 636.47.17 Telex : 241.446 STA UK
STUART, P.	IRISH INDUSTRIAL DEVELOPMENT AUTHORITY Lamsdowne House Lamsdowne Road DUBLIN 4 Ireland Tel. : (1) 688.444 Telex : DUBLIN 4525
TASCHINI, A.	ENTE NAZIONALE PER L'ENERGIA ELETTRICA Bastioni di Porta Volta, 10 20121 MILANO Italy Tel. : (2) 8847.8254 Telex : ENEL-MI 310.496
TREBLE, F.	43, Pierrefondes Ave. Farnborough HANTS GUI4 8PA Great Brtiain Tel. : (252) 42.037
VACELET, G.	S.N.I. AEROSPATIALE 316 Route de Bayonne 31060 TOULOUSE CEDEX France Tel. : (61) 49.11.22. ext. 4079 Telex : 531.546 F
VAN GYSEL, M.	I.D.E. Zoning Industriel 5430 ROCHEFORT Belgium Tel. : (84) 213.771 Telex : 23469 SOLEIL B
VAN OVERSTRAETEN, R.	KATHOLIEKE UNIVERSITEIT LEUVEN ESAT Laboratorium Kardinal Mercierlaan 94 3030 HEVERLEE Tel. : (16) 220.931 Telex : 25941 ELEKUL

VIALART, J.Cl.

PHOTOWATT INTERNATIONAL
98 ter, bd. Héloïse
95.102 ARGENTEUIL
France

Tel. : (1) 982.09.13
Telex : 698.850 F

VISENTIN, R.

UNIVERSITY OF CALABRIA
(PFE - CNR)
P.O. Box 379 COSENZA
Via Nizza, 182
ROMA
Italy

Tel. : (984) 838.139
839.480
(6) 865.493
844.0025
Telex : 800.044 UNICAL I
610.076 CNR RM I
612.322 CNR PFE I

VYVERMAN, A.

SCIENCE POLICY OFFICE
Wetenschapsstraat 8
1040 BRUSSELS
Belgium

Tel. : (2) 236.086
Telex : PROSCIENT BRU B 24501

WILLMES, H.

VARTA BATTERIE AG
Dieckstr. 42
5800 HAGEN
Germany

Tel. : (2374) 393.331
Telex : 082.3841

WOLFE, Ph.

LUCAS ENERGY SYSTEMS LTD.
Windmill Road
Haddenham, Aylesbury
BUCKS, HP1T 8JB
Great Britain

Tel. : (844) 291.681
Telex : 83165 LSO HAD

WHISNANT, R.A.

RESEARCH TRIANGLE INSTITUTE
P.O. Box 12194
RESEARCH TRIANGLE PARK, NC
U.S.A.

Tel. : (919) 541.5825

WRIXON, G.

UNIVERSITY COLLEGE
CORK
Ireland

Tel. : (21) 508.375
Telex : 6050

ZANI, P.E.

ANSALDO
Via N. Lorenzi, 8
16100 GENOVA-CAMPI
Italy

Tel. : (10) 4105.488
Telex : 270.098 ANSLCA

ZOLLENKOPF, K.

HAMBURGISCHE ELECTRICITÄTS-WERKE
Überseering 12
2000 HAMBURG 60

Tel. : (40) 636.31.53
Telex : 217.4121 HEW D

ZWIBEL, H.

NEW MEXICO SOLAR ENERGY INSTITUTE
P.O. Box 390L
LAS CRUCES, New Mexico, 88003
U.S.A.

Tel. : (505) 646.3948

INDEX OF AUTHORS

GPSR Compliance

The European Union's (EU) General Product Safety Regulation (GPSR) is a set of rules that requires consumer products to be safe and our obligations to ensure this.

If you have any concerns about our products, you can contact us on ProductSafety@springernature.com

In case Publisher is established outside the EU, the EU authorized representative is:

Springer Nature Customer Service Center GmbH
Europaplatz 3
69115 Heidelberg, Germany

Batch number: 10614746

Printed by Printforce, the Netherlands